Applied modelling and computing in soc

Janez Povh (ed.)

Applied modelling and computing in social sciences

Bibliographic Information published by the Deutsche Nationalbibliothek
The Deutsche Nationalbibliothek lists this publication in the Deutsche Nationalbibliografie; detailed bibliographic data is available in the internet at http://dnb.d-nb.de.

Library of Congress Cataloging-in-Publication Data

Applied modelling and computing in social sciences / Janez Povh (ed.).
    pages cm
  ISBN 978-3-631-66366-0
  1. Simulation methods. 2. Computer simulation. 3. Industrial management–Simulation methods. I. Povh, Janez, 1973-
  T57.62.A656 2015
  003'.3–dc23
                            2015012901

ISBN 978-3-631-66366-0 (Print)
E-ISBN 978-3-653-05821-5 (E-Book)
DOI 10.3726/ 978-3-653-05821-5

© Peter Lang GmbH
Internationaler Verlag der Wissenschaften
Frankfurt am Main 2015
All rights reserved.
PL Academic Research is an Imprint of Peter Lang GmbH.

Peter Lang – Frankfurt am Main · Bern · Bruxelles · New York ·
Oxford · Warszawa · Wien

This publication has been peer reviewed.

www.peterlang.com

Acknowledgments

This publication is funded by the Creative Core FISNM-3330-13-500033 'Simulations' project funded by The European Regional Development Fund of the European Union. The operation is carried out within the framework of the Operational Programme for Strengthening Regional Development Potentials for the period 2007–2013, Development Priority 1: Competitiveness and research excellence, Priority Guideline 1.1: Improving the competitive skills and research excellence.

Fakulteta za informacijske študije

Faculty of information studies

**Kreativno jedro:
Simulacije**
Creative core: Simulations

»Operacijo delno financira Evropska unija in sicer iz Evropskega sklada za regionalni razvoj. Operacija se izvaja v okviru Operativnega programa krepitve regionalnih razvojnih potencialov za obdobje 2007-2013, 1. razvojne prioritete: Konkurenčnost podjetij in raziskovalna odličnost, prednostne usmeritve 1.1: Izboljšanje konkurenčnih sposobnosti podjetij in raziskovalna odličnost.«

"The operation is partially financed by the European Union, mostly from the European Regional Development Fund. Operation is performed in the context of the Operational program for the strengthening regional development potentials for the period 2007-2013, 1st development priorities: Competitiveness of the companies and research excellence, priority aim 1.1: Improvement of the competitive capabilities of the companies and research excellence."

REPUBLIKA SLOVENIJA
MINISTRSTVO ZA IZOBRAŽEVANJE, ZNANOST IN ŠPORT

Naložba v vašo prihodnost
OPERACIJO DELNO FINANCIRA EVROPSKA UNIJA
Evropski sklad za regionalni razvoj

Table of contents

Introduction

Explaining and predicting social phenomena is always a challenging task.

In this book, we present a wide range of successful applications of modelling and computing in the social sciences. Several authors contributed papers with recent research results. Business process simulation and optimisation, modelling security and weakly defined organisations, optimisation of synergies between different media in marketing, positive and negative sentiment extraction from web media and its analysis over time and media, and medical data-mining are all examples where advanced modelling and computing contributed new knowledge to the social sciences.

Therefore, this monograph demonstrates the power of modelling and computing on one side and the necessity to combine qualitative and quantitative methods to get a complete picture of given social aspect on the other side.

Biljana Mileva Boshkoska, Nadja Damij
Faculty of Information Studies
Sevno 13, 8000 Novo mesto, Slovenia
[biljana.mileva, nadja.damij]@fis.unm.si

An outline for business process modelling using spreadsheets

Abstract: Process simulation is the first step towards the redesign of business process models. However, the leading process simulation tools are usually expensive and often unavailable for common users. On the other hand, almost every computer user has experienced the usage of spreadsheets at least once. Their vast usage is due to its simplicity and to its availability in companies. This paper presents an approach for the usage of spreadsheets for process simulation. We argue that for modelling of simple business processes, the usage of spreadsheets provides just as reliable results as any other expensive simulation tool.

Keywords: business process modelling, spreadsheets

1 Introduction

Business process simulation (BSP) has an essential role in the process of improvement of business process management (BPM) in the public and private sector. BPS is a tool which companies use to get feedback regarding their performances under different conditions and thus enables the redesign of the business process. Regardless of the fact that BPS is acknowledged as relevant and highly applicable, the use of simulation is limited in reality (Nakatumba, Rozinat and Russell 2008). We have managed to observe two main reasons.

The first reason is that there are more than 100 available tools for BPS, each one with different specifics; however, there is a lack of guidance on how to choose a particular tool for simulating a specific task. The second reason is that most of these tools are either too expensive to buy for small companies and start-ups or require a great knowledge for their set up. This paper provides an alternative to the expensive simulation tools by providing an outline for using spreadsheets for simulation of business processes. The reason that we have chosen spreadsheets is because of their vast usage today. They are easy to install and do not require high programming experience to use them.

2 Business Process Modelling and Simulations

In general, a process consists of interconnected activities that require certain resources for their implementation. Thus, to understand the real processes, one has to observe their behaviour and consequently the behaviour of activities and process resources. The observation is usually performed in three steps. Firstly, the process is modelled using some well-known technique. Next, the model is validated in order to find out how close the model reflects the real process. Then the model is analysed by using "what-if" questions to test the options of interest and the functional possibilities functioning of the process. Finally, the model is used for simulation of different scenarios in order to observe the behaviour of the process. The simulations are based on different scenarios using a set of data and assumptions about the process activities and resources (Banks et al. 2001).

Business processes (BP) are modelled with the aim of analysing their current states within the organisation, as well as improving them through the execution of potential "what-if" simulation scenarios (Saven, Olhager 2002). The use of scenario-based what-if analyses enables the design team to test various alternatives and choose the best one (Laguna, Marklund 2005). BP models are simulated with discrete-event simulation (DES) tools. When using DES, the state variable changes only at a discrete set of points in time. When using DES for BPM, the following basic elements have to be considered:

- State: A collection of variables that contain all the information necessary to describe the system at any time;
- Entity: Any object or component of the system which requires explicit representation in the model (i.e. a server, a customer, a machine);
- Attribute: The properties of a given entity (i.e. the priority of a waiting customer, the routing of a job through a job shop);
- Activity: A duration of time of specified length (i.e. a service time or arrival time), which is known when it begins (though it may be defined in terms of a statistical distribution);
- Delay: A duration of time of unspecified indefinite length, which is not known until it ends (i.e. a customer's delay in a last-in, first-out waiting line which, when it begins, depends on future arrivals);
- Clock: A variable representing simulated time.

All of the above described elements can be modelled using spreadsheets.

3 Process Simulation using spreadsheets

3.1 An overview of process modelling using spreadsheets

There are many research papers that describe modelling and simulations using spreadsheets. For example, spreadsheets are used for simulations in healthcare (Klein, Reinhardt 2012; Hartvigsen 2001), in pharmacokinetic processes (Meineke, Brockmöller 2007), and for the simulation of optimisation processes of control parameters (Papa, Mrak 2010). Many companies use spreadsheets for Monte Carlo simulations (Croll 2012; Emmett, Goldman 2004) or optimisation purposes (Croll 2012). However, spreadsheets are used for the simulation of complex physical phenomena, such as the calculation and simulation of the magnetic field of solenoids (Derby, Olbert 2010), or for finite difference time domain simulations for propagating electromagnetic waves (Ward, Nelson 2005), and for the simulation of microprocessor systems (El-Hajj et al. 2000). They are used for the simulation of functions by simulating basic blocks that are interconnected into a graphical interface (El-Hajj 1999) and for the simulation of the work of components of an analog computer so that a simulation of the whole system is obtained (Karim Y. Kabalan 1997). A recent publication in bioinformatics that uses spreadsheets simulates the consequences of varying parameters on measurement results of trace amounts of non-authorised genetically modified organisms (GMO) in feed (Gerdes, Busch, Pecoraro 2014). The processes are usually modelled with equations that are interconnected or by diagrams obtained by interconnecting different shapes offered by the spreadsheets.

3.2 Process Simulation using spreadsheets

Each of the basic DES elements described above, a state, entity, attribute, activity, delay, and clock can be modelled using spreadsheets. We first start with defining the following three elements in spreadsheets: an activity, a transaction, and a gate.

An *activity* has a duration and requires resources for its execution. *Transactions* are flow units that move through the process model from one activity to the next one (iGrafx Help system, 2013). For example, in a model that describes the healthcare process, patients are an example of transactions. They represent entities from the aspect of the defined elements in DES. A *gate* is a modelling element that is activated at a particular time of the day. It means that the transactions that arrive at the gate have to wait until the gate's door opens at the defined time. It represents a *certain delay* in the sense of DES. Uncertain delays are described later. A gate does not have time duration nor resources. Therefore, it is not considered as an activity.

Each modelling element is defined in one spreadsheet. Each row in these spreadsheets represents the *state* of the modelling element at a particular moment of time. A process in spreadsheets is defined by interconnecting the spreadsheets with formulas that lead to the simulation of transactions that move from one modelling element to the next one. The configuration data for each modelling element, such as duration time, and all initial set ups, such as number of starting transactions, and their interconnected start up time are given in a separate worksheet called the *Input Data* worksheet. One example of an *Input Data* worksheet is given in Table 1. Table 1 refers to a process model of the abdominal surgery clinic in Ljubljana, Slovenia. All consecutive tables in this paper refer to the same healthcare process.

Table 1: Excel simulation input data worksheet

Table 2a

Number of beds	75

Table 2b

Parameters of triangular distribution	Activity 18	Activity 24
a	1	3
b	5	5
c	4	4

Table 2c

Activity ID	Activity Name	Resource	Duration (Minutes)
01	Confirm surgery date	1	10
02	Patient reception	1	15
03	View and organize documents	1	10
04	Fulfill care documents	1	15
05	Explain surgery	1	10
06	Sign surgery documents	1	5
07	Sign anaesthesia documents	1	5
08	Prescribe medications	1	10
Gate1			Every morning
09	Prepare patient for surgery	1	30
10	Carry out anaesthesia	1	Between(30;45)
11	Carry out surgery	1	Between(60;240)
12	Wake up patient	1	30
13	Place in intensive care	1	30
14	Prescribe therapy	1	10
15	Addition-al tests	1	10
16	Order additional tests	3	10
17	Carry out care	1	Between(90;180)
18	Check recovery	2	TringDist(1;5;4)
IntG			Every morning
19	Place in Clinic	1	20
20	Prescribe therapy	2	10
21	Addition-al tests	1	10
22	Order additional tests	2	10
23	Carry out care	1	Between(90;240)
ClIG			Every morning
24	Check recovery	2	TringDist(3;5;4)
Gate2			Every morning
25	Inform about release	1	10
26	Prepare release report	1	30
27	Organize transport	1	10
28	Need transport	1	10

The *Input Data* organises the configuration parameters into three tables. The first table, Table 1a, defines the constraints of the model, which for the simulation model of abdominal surgery clinic is presented with the maximum number of available beds. The second table, Table 1b, provides various data about the parameters of the modelled activities, described with different distributions.

Table 1b gives the parameters of the triangular distributions that are required in order to determine the duration of some of the activities, such as activities 18, 24, and others. For example, a triangular distribution determines the duration of the patient's stay in intensive care (activity 18 in Table 1c). It has the following parameters: minimum stay of the patient in the intensive care is one day, maximum stay is five days, and the most probable stay is four days. The actual duration is estimated with the function "rand()" that generates a pseudo-random number uniformly distributed in the interval (0,1), which is scaled to the desired distribution according to the given parameters in Table 1b. This transformation is known as a Probability Integral Transformation (Leemis, Park 2005). This way, we model a *delay* from the aspect of the mentioned DES elements.

The last table, Table 1c, shows the *attributes* for each activity, such as ID, number of resources, and duration.

Each of the spreadsheets modelling elements is described in a separate worksheet. The activities worksheets start with a column that states the actual time in the simulation. This column is a model of the *clock* in the DES modelling elements.

3.3 Types of modelling elements

Each of the three modelling elements – gate, activity, and transaction is modelled with a separate spreadsheet. Each row in the spreadsheet represents the state of the modelling element at a particular moment of time.

3.3.1 Gate

The model of a gate in a worksheet comprises the following four columns:

(A)   states the time when the gate is activated;
(B)   gives the number of transactions that are currently waiting at the gate;
(C)   lists all transactions that wait at the gate; and
(D)   lists all transactions which are ready to be propagated to the next activity.

An example of a gate is presented in Figure 1. The first column is the *certain delay* that shows the activation time moment of the gate element.

Biljana Mileva Boshkoska, Nadja Damij

Figure 1: Example of a gate modelling element: Gate1

A	B	C	D
Time in minutes	Patients waiting	All transactons that arrived	Transactions waiting at 8:00 AM
481	25	1:2:3:4:5:6:7:8:9:10:11:12:13:14:15:16:17:18:19:20:21:22:23:24:25	1:2:3:4:5:6:7:8:9:10:11:12:13:14:15:16:17:18:19:20:21:22:23:24:25
961	0	1:2:3:4:5:6:7:8:9:10:11:12:13:14:15:16:17:18:19:20:21:22:23:24:25:26	26:27:28:29:30:31:32:33:34:35:36:37:38:39:40:41:42:43:44:45:46:47:48:49:50:51:52:53:54:55
1441	0	1:2:3:4:5:6:7:8:9:10:11:12:13:14:15:16:17:18:19:20:21:22:23:24:25:26	56:57:58:59:60:61:62:63:64:65:66:67:68:69:70:71:72:73:74:75

3.3.2 Activity

Figure 2 presents an example of an activity. The worksheet that models an activity is comprised of the following eight columns (A-H):

(A) States the simulation minute of the time which applies to the whole row in the worksheet;

(B) Stores whether a transaction has arrived in the indicated minute; "one" denotes arrival of a transaction, and "zero" denotes that no transaction has arrived;

(C) Shows until when a resource is busy; meaning until which minute;

(D) States the availability of the resource; "one" denotes that the resource is busy, and "zero" denotes that the resource is available;

(E) Gives a cumulative number of transactions that are waiting in a queue for processing;

(F) Indicates whether the transaction defined in row I is finished or not. If it is finished then one is inserted in cell (I, F), which means that the current transaction continues in the next activity. Otherwise, "zero" is written in cell (I,F);

(G) Lists all transactions that are waiting in a queue for processing at the current activity. The transaction that will be the first processed is first in the queue.

(H) The last column lists the transactions that were processed by the current activity. Consequently, these activities are propagated to the next worksheet (activity). The values are given as integer numbers, representing the transaction IDs.

(I) When a gate follows an activity, we need to list all transactions that have finished with the activity.

Generally, an activity input is a transaction that has finished its execution from the preceding activity. In continuation we describe, which may have multiple sources of incoming transactions and a decision activates.

Figure 2: Spreadsheet of "Confirm surgery date" activity

	A	B	C	D	E	F	G	H
	Time in minutes	Arrival of transaction	Resource busy until	Resource Busy (1 = busy; 0 = free)	Transactions waiting for execution	Finished transaction	List of transactions that started	Transaction ID ended
1	1	1	11	1	0	0	1	0
2	2	1	11	1	1	0	1:2	0
3	3	1	11	1	2	0	1:2:3	0
4	4	1	11	1	3	0	1:2:3:4	0
5	5	1	11	1	4	0	1:2:3:4:5	0
6	6	1	11	1	5	0	1:2:3:4:5:6	0
7	7	1	11	1	6	0	1:2:3:4:5:6:7	0
8	8	1	11	1	7	0	1:2:3:4:5:6:7:8	0
9	9	1	11	1	8	0	1:2:3:4:5:6:7:8:9	0
10	10	1	11	1	9	0	1:2:3:4:5:6:7:8:9:10	0
11	11	1	11	0	10	1	2:3:4:5:6:7:8:9:10:11	1

3.3.3 Different types of activities

The model distinguishes two main modifications of an activity: a decision activity and activity with multiple sources of incoming transactions. In the latter case, the model defines two modifications of the activity depending on the type of the modelling element that generates the incoming transaction to the activity:

1. one transaction is generated from a preceding activity and another is generated from a preceding decision activity, or
2. one transaction is generated from a preceding activity and another is generated from a preceding gate.

3.3.4 Decision activity

A decision activity is used to check whether a certain condition is fulfilled and on the basis of the outcome, decides for further actions. The decision activity usually has two paths. One path is indicated by "YES" and means that the condition is fulfilled. "NO" denotes the other path.

Figure 3 is an example of a model of a decision activity. In addition to the columns defined for an activity, it comprises the following columns:

(I)   generates a random number using the "RAND()" function;
(J)   gives a decision YES or NO based on the obtained number in (A);
(L)   auxiliary column that keeps record of the transactions that ended with a decision "YES", and
(M) auxiliary column that keeps record of the transactions that ended with a decision "NO".

Figure 3 Example of a decision activity 18

	A	B	C	D	E	F	G	H	I	J	L	M
Row No.	Time in minutes	Arrival of transaction	Resource busy until	Resource Busy (1 = busy; 0 = free)	Transactions waiting for execution	Finished transaction	List of transactions that started	TransactionID ended	Decision	20% YES; 80% NO	Auxiliary column YES; follow the gate INT	Auxiliary column NO; follow the activity 19
163	1316	1	1326	0	0	0	49	0	0,56087	NO	
314	1448	1	1458	0	0	0	74	0	0,88199	YES	49
446	1552	1	1562	0	0	0	99	0	0,010479	NO	49
550	1722	1	1732	0	0	0	124	0	0,605636	NO	49
720	1878	1	1888	0	0	0	49	0	0,661489	NO	49
876	2020	1	2030	0	0	0	149	0	0,233557	NO	49
1178	2178	1	2188	0	0	0	174	0	0,005266	NO	49
1305	2307	1	2317	0	0	0	199	0	0,796961	NO	49
1422	2424	1	2434	0	0	0	224	0	0,51937	NO	49
1577	2579	1	2589	0	0	0	249	0	0,974492	YES	49
1748	2750	1	2760	0	0	0	274	0	0,023592	NO	49
1926	2928	1	2938	0	0	0	299	0	0,518122	NO	49:274

3.3.5 Activity with incoming transactions from two preceding activities

Figure 4 represents the activity 17, which is a successor of two activities: the decision activity 15 and the activity 16. The arrival of transactions is simulated with three columns: columns B and C show arrival of the transaction from activities 15 and 16, respectively, and column D shows if the transaction has arrived from either activity 15 or 16.

The time for execution of this activity is uncertain, in the sense that it may take any value between 90–180 minutes (see activity 17 in Table 1c). Consequently, to simulate the time execution, a random number is generated in the interval (0–1) in each minute, which is scaled to the required interval.

Figure 4 Example of an activity 17 with two inputs: from activity 15 and 16

	A	B	C	D	E	F	H	I	J	M	N
1	Time in minutes	Arrival of transaction from 15	Arrival of transaction from 16	Arrival of transaction from 15 or 16	Resource busy until	Resource Busy (1 = busy; 0 = free)	Transactions waiting for execution	Finished transaction	Uniform distribution	List of transactions that started	Transaction ID ended
2	4804	0	0	0	4888	1	1	0	180	24:21	0
3	4805	0	0	0	4888	1	1	0	109	24:21	0
4	4806	0	0	0	4888	1	1	0	96	24:21	0
5	4807	0	0	0	4888	1	1	0	169	24:21	0
6	4808	0	0	0	4888	1	1	0	95	24:21	0
7	4809	1	0	1	4888	1	2	0	170	24:21:18	0
8	4810	0	0	0	4888	1	2	0	155	24:21:18	0
9	4811	0	0	0	4888	1	2	0	138	24:21:18	0
10	4812	0	0	0	4888	1	2	0	159	24:21:18	0

3.3.6 Activity with incoming transactions from preceding activity and a gate

The second modification of an activity with multiple incoming transactions discusses the case when the incoming transactions come from: a preceding gate and a preceding activity. Figure 5 shows Activity 14 as an example of this case. Columns H and I are the newly introduced ones in the model of such an activity. Column H checks whether it is time to remove transactions from the Gate *IntG*.

Figure 5 Example of an activity 14 with two inputs: from activity ID 13 and the gate activity ID IntG

A	B	C	D	E	F	H	I	J	M	N
Time in minutes	Arrival of transaction	Resource busy until	Resource Busy (1 = busy; 0 = free)	Transactions waiting for execution	Finished transaction	Check if time to withdrow transactions from IntG	Transaction has arrived from IntG	Transaction from 13 or from IntG	List of transactions that started	TransactionID ended
726	1	736	1	0	0	0	0	1	1	0
960	0	938	0	0	0	1	0	0		0
1440	0	1352	0	1	0	1	3	1	3	0
1920	0	1801	0	0	0	1	0	0		0
2400	0	2354	0	0	0	1	0	0		0
2880	0	2799	0	3	0	1	10:11:12	1	10:11:12	0
3360	0	3345	0	2	0	1	13:14	1	13:14	0
3840	0	3810	0	2	0	1	12:16	1	12:16	0
4320	0	4296	0	1	0	1	17	1	17	0

3.3 Outputs of the model

The output of the simulation model provides calculations and graphical representation of distributions of the following parameters:

1. Minimum cycle time
2. Average cycle time
3. Maximum cycle time
4. Average work time
5. Average wait time

Work time is the simulated amount of time that activities are processed. It is directly calculated as a sum of the duration times of the activities.

Wait time is the simulated amount of time that transactions are waiting. The waiting may occur due to different reasons such as waiting for a resource or because of a gate.

Cycle time is the simulated amount of time that a transaction spends within the process starting with the first activity and ending with the last one. A cycle time may differ from one activity to another due to the different paths in the process and to the different activity durations calculated from their distributions. The cycle time is a sum of work time and wait time.

The following equation connects the last three parameters:

Cycle time = Work time + Wait time.

Maximum (minimum) Cycle time refers to the highest (lowest) accumulated time for the transactions at hand, at any point in time.

The term *average* refers to the total time that all transactions spent in each activity divided by the number of completed transactions for each activity.

4 Conclusion

Simulation is a powerful tool for process analysis because there are currently more than 100 different software tools developed for simulation. The main difficulty that practitioners of BP face is choosing the right simulation tool for a given business process because of two reasons: the prize of the offered tools and the limited literature on comparison of these tools. In this paper, we propose an outline for using spreadsheets for discrete event simulation of business process models. Spreadsheets are widely accessible to practically all computer users and also require little programming knowledge; however, they require a suitable initial time to build the model. We propose for the usage of spreadsheets in cases with a medium complexity of processes and financial limitations.

5 Acknowledgements

This work is supported by Creative Core FISNM-3330–13-500033 'Simulations' project funded by The European Regional Development Fund of the European Union.

References

Banks, Jerry / John S. Careson / Nelson Barry L. / Nicol David M.: *Discrete-Event System Simulation*. New Jersey: Prentice-Hall, Inc. 2001.

Croll, Grenville J.: "Spreadsheets and Long Term Corporate Survival." *Proceedings of European Spreadsheet Risks Interest Group* 2012, pp. 77–96.

David Hartvigsen Mendoza: *SimQuick Process Simulation with Excel*. College of Business Administration, University of Notre Dame. Prentice Hall, Upper Saddle River, NJ 07458. 2001.

Derby, Norman / Stanislaw Olbert: "Cylindrical magnets and ideal solenoids." *American Journal of Physics* 78 (3), 2010, pp. 229–235.

El-Hajj, Ali: " Functional Simulation Using Spreadsheets." *Simulation* 73 (2), 1999, pp. 80–90.

El-Hajj, Ali / Karim Y. Kabalan / Maher N. Mneimneh / Feras Karablieh: "Microprocessor Simulation and Program Assembling Using Spreadsheets." *Simulation* 75 (2), 2000, pp. 82–90.

Emmett, Hilary L. / Lawrence I. Goldman: "Identification of Logical Errors through Monte-Carlo Simulation." *Proc. European Spreadsheet Risks Int. Grp. (EuSpRIG)*. 2004.

Gerdes, Lars/ Ulrich, Busch / Sven, Pecoraro: "A statistical approach to quantification of genetically modified organisms (GMO) using frequency distributions." *BMC Bioinformatics* 15 (1), 2014, pp. 1–12.

Hartvigsen, David: *SimQuick Process Simulation with Excel.* Mendoza College of Business Administration, University of Notre Dame, Prentice Hall, Upper Saddle River, NJ 07458. 2001.

2011. "http://www.igrafx.com/company/news/article/igrafx-spin-off-establishes-leading-independent-provider-of-business-process-analysis-solutions." Accessed January 20, 2015.

2013. *iGrafx Help system.* iGrafx LLC.

Kadjo, Akinde / Purnendu K. Dasgupta: "Tutorial: simulating chromatography with Microsoft Excel Macros." *Anal Chim Acta.* 773, 2013, pp. 1–8.

Karim Y. Kabalan / Ali El-Hajj / Hassan Diab / Souhier Fakhreddine: "Analog Computer Simulation Using Spreadsheets." *Simulation* 68 (2), 1997, pp. 101–106.

Klein, Michael G. / Gilles, Reinhardt: "Emergency Department Patient Flow Simulations Using Spreadsheets." *Simulation in Healthcare* 7 (1), 2012, pp. 40–47.

Laguna, Manuel / Marklund, Johan: *Business Process Modeling, Simulation, and Design.* New Jersey: Pearson Education, Inc. 2005

Leemis, Lawrence / Steve Park: *Discrete-event simulation: a first course.* Upper Saddle River, NJ, USA: Prentice-Hall, Inc. 2005.

Macho, Siegfried: "Cognitive modeling with spreadsheets." *Behav Res Methods Instrum Comput* 34 (1), 2002, pp. 19–36.

Meineke, Ingolf / Jürgen Brockmöller: "Simulation of complex pharmacokinetic models in Microsoft EXCEL." *Computer Methods and Programs in Biomedicine* 88 (3), 2007, pp. 239–245.

Mileva Boshkoska, Biljana / Damij, Talib / Jelenc, Franc / Damij, Nadja: "Abdominal Surgery Process Modeling Framework for Simulation Using Spreadsheets." *Computer Methods and Programs in Biomedicine.* 2015 (submitted).

Nakatumba, Joyce / Rozinat, Anne / Russell, Nick: "Business Process Simulation: How to get it right?" In *International Handbook on Business Process Management.* Springer-Verlag. 2008.

Papa, Gregor / Peter Mrak: "Optimization of cooling appliance control parameters." *2nd International Conference on Engineering Optimization.* Lisbon, Portugal. 2010.

Saven, Ruth Sara / Olhager, Jan: "Integration of Product, Process and Functional Orientations: Principles and a Case Study." *Preprints of the International Conference on Advanced Production Management Systems, APMS, IFIP.* The Netherlands. 2002.

SN, Wu: "Simulations of the cardiac action potential based on the Hodgkin-Huxley kinetics." *Chinese Journal of Physiology* 47 (1), 2004, pp. 15–22.

Ward, David W. / Keith A. Nelson: "Finite Difference Time Domain (FDTD) Simulations of Electromagnetic Wave Propagation Using a Spreadsheet." *Computer Applications in Engineering Education* 13 (3), 2005, pp. 213–221.

Yusuf, Jafry / Fredrika, Sidoroff / Roger, Chi: "A Computational Framework for the Near Elimination of Spreadsheet Risk." *Proc. European Spreadsheet Risks Int. Grp. (EuSpRIG)*, 2006, pp. 85–89.

Tadej Kanduč, Blaž Rodič
Faculty of Information Studies
Sevno 13, 8000 Novo mesto, Slovenia
{tadej.kanduc, blaz.rodic}@fis.unm.si

Manufacturing processes optimisation in a furniture factory

Abstract: In this paper, we present a project of optimising the manufacturing processes in a Slovenian furniture development company. The key steps of the project, such as extracting and preparing data, synchronising different software environments, building a simulation model, and optimising processes are outlined in this paper.
We have constructed a discrete event simulation (DES) model of manufacturing processes, which allows us to understand the current process and optimise its parameters. We have also developed a method for automated model construction which modifies the simulation model according to the input parameters. At the end, we present the optimisation problem handled in the project: minimisation of the total distance the products need to travel between the machines. The problem was solved by optimising the machine layout on the factory floor using our newly developed heuristic method.

Keywords: discrete event simulation, optimal layout, heuristic optimisation

1 Introduction

Manufacturing processes in larger production companies are generally complex and need to be systematically organised in order to achieve high levels of efficiency. Companies need to consider several criteria and restrictions in the processes such as costs, due dates, amounts of stock materials, different measurements in efficiency, etc.

Simple manufacturing processes are usually formalised using mathematical language and are optimised with exact mathematical methods. On the other hand, discrete event simulation (DES) modelling is more suitable to describe a complex manufacturing system. Construction of a DES model requires that the data that describes the manufacturing processes are obtained, analysed, extracted, and prepared in a suitable format for the model. System or process optimisation can be performed by implementing changes in the model, usually by constructing several versions of the model and input data (i.e. scenarios) and comparing simulation results. If a large number of model variations needs to be built, it is convenient to design a tool that constructs all the models automatically according to defined input parameters.

In the paper, we present the main steps of the project of optimising manufacturing processes in a Slovenian furniture factory. The primary goal of the company is to reduce overall costs in manufacturing processes. Our goal was to investigate how the layout of machines on the factory floor affects the efficiency of manufacturing processes. Our primary optimisation criterion was the total distance the manufactured products needed to travel on the floor; however, we have also monitored other criteria during the optimisation processes. The results of our project are used within a currently running micro logistics optimisation process.

Figure 1: Segment of the factory floor (machines and routes between them).

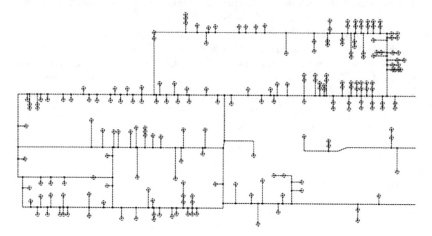

1.1 Problem situation

Approximately 140 machines are located on the factory floor of the company (see Figure *1*). The company catalogue contains more than 30,000 different products. Each product is manufactured according to the prescribed bill of materials (BOM) and its technical procedure. In the BOM, the required semi-finished products and materials to manufacture the given product are listed. The technical procedure data describes the sequence of operations in the manufacturing of the given product. Manufacturing processes include a large set of different products and variations of open orders during each working month. Production scheduling is based on customer orders and performed using the Preactor scheduling system.

1.2 Related work

Simulation is commonly used for the evaluation of scenarios (Kljajić/Bernik/ Škraba 2000; Edis et al. 2011; Tearwattanarattikal/Namphacharoen/Chamrasporn 2008). However, the models developed with the visual interactive modelling method (VIM) are usually manually constructed through careful analysis of the real-life system and communication with process owners. Automated model development is more common with methods that allow easier and more standardised formal description of models, i.e.. Petri nets (Conner 1990; Gradišar/Mušič 2012). The automation of model construction and adaptation can importantly facilitate the development of models of complex systems (Lattner et al. 2010; Kannan/Santhi 2013) and the generation of simulation scenarios.

Several papers deal with factory layout optimisation. Benjaafar/Heragu/Irani (2002) states that multiproduct enterprises require a new generation of factory layouts that are flexible, modular, and easy to reconfigure. Evolutionary optimisation methods are often proposed due to problem complexity (Sadrzadeh 2012). The layout optimisation problem is identified as hard Combinatorial Optimisation Problem and the Simulated Annealing (SA) meta-heuristic resolution approach is proposed to solve the problem (Moslemipour/Lee 2012). A novel particle swarm optimisation method is proposed by (Ficko et al. 2010) for the intelligent design of an unconstrained layout in flexible manufacturing systems.

Factory layout design optimisation is further discussed in Krishnan/Cheraghi/Nayak (2008), Kochhar/Heragu (1999), and Enea/Galante/Panascia (2005). Krishnan/Cheraghi/Nayak (2008) proposes a new facility layout design model to optimise material handling costs. Kochhar/Heragu (1999) and Enea/Galante/Panascia (2005) propose genetic algorithm based solutions to respond to the changes in product design, mix, and volume in a continuously evolving work environment.

The optimisation of the floor layout to minimise the total product distance is closely related to the so-called quadratic assignment problem. The latter problem was firstly introduced in Koopmans/Beckmann (1957) and the problem was proven to be NP-hard (Sahni/Gonzalez 1976). Larger problems need to be solved using heuristic methods since exact algorithms are ineffective. There are several available open source quadratic assignment problem (QAP) heuristic method implementations (see Taillard (1998), Shah, Tsourapas (2008), and QAPLIB website (http://www.opt.math.tu-graz.ac.at/qaplib/codes.html)).

Force-directed methods are one of the most commonly used methods for drawing graphs. The widespread use comes from their simplicity and visually appealing representation of the graphs (see Fruchterman/Reingold (1991), Kobourov (2013), and references therein).

2 Methodology

Our task in the project was to develop a better machine layout, which will fit the current production needs and projections for the next ten years. To complete this task, we have developed a simulation model of the factory floor and optimisation methods based on the company data and their specific optimisation goals. The purpose of the model is for the verification of new manually or algorithmically generated floor layouts. For most of the machines, there are no specific location restrictions. It is also be possible to add some new machines on the floor if considerable improvements can be achieved.

The optimisation of the floor layout is conducted in cooperation with experienced manufacturing planners, managers, and other experts within the company, and is facilitated by state-of-the-art optimisation algorithms that are employed to generate new layout scenarios (i.e. to search for the optimal layout within a large set of possible layouts).

2.1 Existing tools and data

As part of an established scheduling and planning procedure, Preactor software (http://www.preactor.com/) is used to schedule customer orders. It monitors daily manufacturing processes on the factory floor and suggests scheduling plans according to a set of priorities and the availability of resources (machines). Preactor is a family of "advanced scheduling and planning" products that allows for a detailed definition of manufacturing and other processes, and integrates with existing ERP and other company databases and applications.

However, the modelling process within Preactor is not flexible enough to allow for the easy modification of the system model or modelled processes and testing of scenarios required for layout or process optimisation. To simulate processes in a different factory floor layout, an entire simulation model needs to be built from scratch or undergo lengthy manual modification.

2.2 Selection of tools and methods

We decided to implement current production processes and the optimisation procedure with a specialised simulation and modelling tool, Anylogic, a powerful software that implements DES, system dynamics (SD), and agent based modelling methodologies. Modelling is performed using the VIM approach, which is intuitive and clear and supports advanced visualisation techniques. Anylogic or other simulation and modelling tools are not a replacement for advanced scheduling and planning tools such as Preactor or vice versa. Instead, they complement each other: Preactor contains a detailed process model that allows for accurate scheduling and planning and provides detailed process data for Anylogic. On the other hand, in Anylogic, designing, processes optimisation, adding new machines, and verifying the scenarios using different factory layouts and sets of orders is easy and intuitive.

The Anylogic simulation model allows us to monitor various manufacturing process statistics and to better understand the manufacturing system by discovering rules and connections in the manufacturing system. The model was verified by comparing the simulation results (i.e. manufacturing time, machine utilisation) using synthetic and real historic order data prepared by the company planners with the real-event statistics of the set of orders from the past year.

An important part of the project was the preparation and exportation of manufacturing process data and customer order data from the company database. Synchronisation of all software components (databases, simulation model, model construction application, and auxiliary applications) in order to achieve the required level of integration was also needed. Part of the data preparation and synchronisation was done in our external Java program.

Anylogic stores the models as standard XML files, which allows easy manual or algorithmic modifications of the model. We have developed an application in a Java environment that reads input data from Excel and template XML simulation model, and constructs the corresponding Anylogic model.

2.3 Simulation model

There are around 140 machines (working places) on the factory floor. Each machine is operated independently and takes up a certain amount of space on the factory floor. Around the machines, a network of routes is defined. We presume that all carts move on these paths with constant speed and use the shortest route possible.

Each machine can perform specific operations. Before starting an operation, the machine needs to be set up to be able to perform the prescribed tasks. Both the set up and operation of the machine take up a specific amount of time.

A machine has the input and the output pallet (a landfill for the products). Once the operations for the series of products are finished, a cart moves the series of products to the next chosen machine. Every machine is described by the following procedure:

(1) If input pallet is not empty, pick a product among the list; else wait.
(2) Set up the machine.

(3) Apply the operations for the whole series of the product (products then wait on the output pallets).
(4) Pick a suitable machine for the next operation that is the least loaded at that moment.
(5) Series of product is moved to the next machine.
(6) Go to (1).

The manufacturing process is presented as a DES model in Anylogic. The structure representing a machine is the most important and complex segment of the model. It consists of several blocks: input and output blocks, arrays to monitor products on the pallets, machine delay blocks, queues, etc. There are also several action/ event blocks that are activated when new products come to the input pallet, the machine idles, the machine is setting up, etc. Machines are connected using the actual network of routes on the factory floor. Products and carts are represented as transactions between the machines.

2.4 Data preparation for the simulation model

The manufacturing process data includes the information for technical procedures and the BOM. They are stored in a Microsoft SQL Server database, used mainly by the manufacturing scheduling application Preactor. We have analysed the structure and content of database tables and prepared a set of queries that were used to extract the data required for the model construction and simulation scenario generation (i.e. the preparation of model input data). The queries were stored in the Microsoft SQL Server database in the form of views and later called by an Excel workbook. The workbook was used as an intermediate data storage mechanism that allowed us to examine and modify the data as required. Some corrections were necessary as the original database contained some errors and some data was missing for certain technical procedures and the BOM. This is an inevitable step when dealing with real-life data. Table 1 shows an example of an SQL query used to obtain the data on machines in machine groups (referred to as Resources and ResourceGroups).

Table 1: Example of an SQL query.

```
/*Podgorje_baza_20140403.LSI.*/
CREATE VIEW Test19Projects_equivMachines AS
        SELECT ResourceGroupId, RGR.ResourceId, ResourceCode FROM
               Podgorje_baza_20140403.LSI.ResourceGroupResources RGR,
               Podgorje_baza_20140403.LSI.Resources R
        WHERE RGR.ResourceId=R.ResourceId
;
```

2.5 Input and output simulation data

All the input data (orders, technical procedures, BOM, etc.) are primarily stored in the SQL databases generated by the Preactor software. Relevant data are saved as queries and exported to an intermediate Excel file. In Excel, the data are manually slightly modified since inaccuracies and inconsistencies in real data occasionally occur. In Excel, the following input data are stored:

- An order is described as a list of products (catalogue numbers). For every product from the list, name, quantity, earliest start time, priority parameter, and volume are assigned.
- Each product has a specific technical procedure. For every operation, there is a group of equivalent machines, a preferred machine, set up time, and time per item.
- More complex products also have the BOM (i.e. list of required semi-finished products or materials that are joined at a specific operation in specific quantity).

At the start-up of the simulation, the input data from Excel are read and stored in internal Anylogic arrays. From there on, all data are read from internal data structures to remove constant communication with external files, which would slow down the simulation.

During simulation, various statistical data are measured and stored:

- For every pair of machines, different types of flows (number of products, number of used carts, overall volume of products, and total distance of carts) are measured.
- For every machine, utilisation, overall set up time, flow of products and volume, and queue of products are monitored.
- For each series of products, completion times and sequences of machines, which were chosen during simulation, are stored.
- Different and less significant measurements, such as flow of carts and routes of the carts, are recorded.

Once the simulation is finished, all the data are stored in the output Excel file.

2.6 Components of the simulation system

The modelling and simulation system is composed of four main elements:

- Core manufacturing process simulation model in the Anylogic environment.
- Java application that constructs the XML Anylogic model from a template file.
- MS Excel as an intermediate input and output data storage and analysis tool.

- MS SQL server database describing technical procedures and client's orders.

The resulting system is shown in Figure 2. The simulation run is prepared as follows. First, we prepare the Anylogic template file (XML). The simulation model (new Anylogic XML file) is constructed by running the Java algorithm for automatic model building. Next, we run the Anylogic simulation model. During simulation, the input database is read dynamically. When simulation is finished, simulation results are stored in an output Excel file.

Figure 2: System schematics.

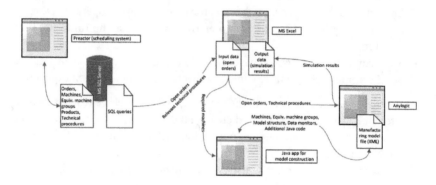

2.7 Optimisation methods

In this section, we present the factory floor optimisation problem. The goal is to minimise the total transportation distances of the products during production. We have tried different optimisation algorithms to minimise the total distance of the products. We have tested freely available open source heuristic algorithms in C++ and Matlab for the quadratic assignment problem that are based on simulated annealing (Taillard 1998), iterative local approach (Shah), and ant colony algorithm (Tsourapas 2008). We have also designed a heuristic optimisation method using the SD methodology in Anylogic that is based on force-directed drawing algorithms. So far, it has generated good results and will be further developed and explored.

The optimisation problem is presented as finding the optimal mathematical network, in which nodes of the network represent the machines on the factory floor and weighted edges between the nodes represent transactions between the machines. Real routes on the floor between the machines are neglected in this case, since it considerably complicates the optimisation problem. The optimisation method should only propose a basic outline of the layout, since the final layout needs to be further tuned by the company experts to meet other less precise criteria.

Each machine takes a specific amount of space on the floor and the machine regions must not intersect. If we further presume that all machines take the same amount of space on the floor, we can restrict the machine positions to discrete points on a predefined grid. Hence the problem simplifies to well-known QAP.

As an alternative to QAP algorithms, we have developed a promising alternative optimisation method, which is based on force-directed graph drawing methods. We define attractive and repelling forces between the machines. In the simulation process, the machines are repositioned according to the defined forces in the system to achieve local minimum energy of the system. When the machines do not move any more, the optimisation process is finished.

To every machine (m_i), we prescribe the corresponding repelling force (F_{ij}) to every other machine (m_j). Repelling forces keep the machines away from each other since we want sufficient space between the machines. For every pair of machines (m_i, m_j), we define the attractive force (G_{ij}). It is proportional to the amount of direct product volume transactions between m_i and m_j and the distance between m_i and m_j. Attractive forces move the machines with larger volume transactions closer to each other.

3 Results and discussion

The main outcomes of the project are the integrated simulation model in Anylogic that communicates with external database files, the method for automatic model construction, and the novel heuristic floor layout optimisation method. The model servers as an indispensable tool for the in-depth analysis of the manufacturing process.

The novel optimisation method outperformed other more general heuristic methods for QAP in terms of the optimisation criterion for this particular problem. The generated layout has an approximately 20 % shorter total product travel distance than the current layout. Shorter travel also means less time that the workers need to transport the products. The customer has responded very favourably to these results. Furthermore, it has prepared several manually adjusted floor layouts based on our generated layout that will be verified with the simulation model before selection and implementation.

Further steps in our project will include changing the set of machines: replacement of one or several machines by newer multipurpose CNC machines. Other optimisation goals and criteria (changing order priorities, minimising total production time, etc.) will be explored in the future.

4 Acknowledgements

Work supported by Creative Core FISNM-3330-13-500033 'Simulations' project funded by The European Regional Development Fund of the European Union. The operation is carried out within the framework of the Operational Programme for Strengthening Regional Development Potentials for the period 2007–2013, Development Priority 1: Competitiveness and research excellence, Priority Guideline 1.1: Improving the competitive skills and research excellence.

5 References

Benjaafar, Saif / Heragu, Sunderesh S. / Irani, Shahrukh A.: "Next generation factory layouts: Research challenges and recent progress". Interfaces, 2002, 32(6), pp. 58–76.

Conner, William: Automated Petri net modeling of military operations, in IEEE Proceedings of the IEEE 1990 National Aerospace and Electronics Conference – NAECON 1990, Volume 2, 1990, Dayton, Ohio, USA, pp. 624–627.

Edis, Rahime S./ Kahraman, Bayram / Araz, Özlem U. / Özfirat, M. Kemal: "A facility layout problem in a marble factory via simulation". Mathematical and Computational Applications, 2011, 16(1), pp. 97–104.

Enea, M. / Galante G. / Panascia, E.: "The facility layout problem approached using a fuzzy model and a genetic search". Journal of Intelligent Manufacturing, 2005, 16(3), pp. 303–316.

Ficko, Mirko / Brezovnik, Simon / Klancnik, Simon / Balic, Jože / Brezocnik, Miran / Pahole, Ivo: "Intelligent design of an unconstrained layout for a flexible manufacturing system". Neurocomputing, 2010, 73(4–6), pp. 639–647.

Fruchterman, Thomas M. J. / Reingold, Edward M.: "Graph drawing by force-directed placement". Software-Practice and Experience, 1991, 21(11), pp. 1129–1164.

Gradišar, Dejan / Mušič, Gašper: "Automated Petri-Net Modelling for Batch Production Scheduling", in (ed: Pawel Pawlewski) Petri Nets – Manufacturing and Computer Science, InTech, 2012, pp. 3–26. http://dx.doi.org/10.5772/48467

Kannan, Ramesh / Santhi, Helen: "Automated construction layout and simulation of concrete formwork systems using building information modeling". In (eds: Djwantoro Hardjito & Antoni) Proceedings of The 4th International Conference of Euro Asia Civil Engineering Forum 2013 (EACEF 2013), National University of Singapore, 26–27 June 2013, pp C7–C12.

Kljajić, Miroljub / Bernik, Igor / Škraba, Andrej: "Simulation approach to decision assessment in enterprises". Simulation, 2000, 75(4), pp. 199–210.

Kobourov, Stephen G.: "Force-directed drawing algorithms". In (eds: R. Tamassia) Handbook of Graph Drawing and Visualization, CRC Press, 2013, pp. 383–408.

Kochhar, Jasmit S. / Heragu, Sunderesh S.: "Facility layout design in a changing environment". International Journal of Production Research, 1999, 37(11), pp. 2429–2446.

Koopmans, Tjalling C. / Beckmann, Martin: "Assignment problems and the location of economic activities". Econometrica, 1957, 25(1), pp. 53–76.

Krishnan, Krishna K. / Cheraghi, S. Hossein / Nayak, Chandan N.: "Facility layout design for multiple production scenarios in a dynamic environment". International Journal of Industrial and Systems Engineering, 2008, 3(2), pp. 105–133.

Lattner, Andreas D. / Bogon, Tjorben / Lorion, Yann / Timm, Ingo J.: "A knowledge-based approach to automated simulation model adaptation". In Proceedings of the 2010 Spring Simulation Multiconference (SpringSim '10). Society for Computer Simulation International, San Diego, CA, USA, 2010, Article 153.

Moslemipour, Ghorbanali / Lee, T. S.: "Intelligent design of a dynamic machine layout in uncertain environment of flexible manufacturing systems". Journal of Intelligent Manufacturing, 2012, 23(5), pp. 1849–1860.

Sadrzadeh, Amir: "A genetic algorithm with the heuristic procedure to solve the multi-line layout problem". Computers and Industrial Engineering, 2012, 62(4), pp. 1055–1064.

Sahni, Sartaj / Gonzalez, Teofilo: "P-complete approximation problems". Journal of the ACM, 1976, 23(3), pp. 555–565.

Shah, Shalin: "Implementation of iterative local search (ILS) for the quadratic assignment problem", http://shah.freeshell.org/ilsassignment/ cILSAssignment. cpp.

Taillard, Erik: "Simulated annealing (SA) procedure for the quadratic assignment problem", 1998, http://mistic.heig-vd.ch/taillard/codes.dir/sa_qap.cpp.

Tearwattanarattikal, Pochamarn / Namphacharoen, Suwadee / Chamrasporn, Chonthicha: "Using ProModel as a simulation tools to assist plant layout design and planning: Case study plastic packaging factory". Songklanakarin Journal of Science and Technology, 2008, 30(1), pp. 117–123.

Tsourapas, Kyriakos: "Ant Algorithm for the Quadratic Assignment Problem", 2008, http://www.mathworks.com/matlabcentral/fileexchange/1663-qap.

Borut Rončević, Urška Fric
Faculty of Information Studies
Sevno 13, 8000 Novo mesto, Slovenia
{borut.roncevic, urska.fric}@fis.unm.si

Researching Industrial Symbiosis: Challenges and Dilemmas

Abstract: While industrial symbiosis research has recently gained its popularity, research on its social aspects is still in its nascent phase. This paper presents some main challenges and dilemmas that need to be tackled in the future. Industrial symbiosis can be thought of as a relationship between two or more economic actors involved in the exchange of material resources, where each economic actor adds to the welfare of other actors and the society in general, while following their own benefits. Social aspects have a key role in the formation and embeddedness of industrial symbiosis in its natural, industrial, and social settings. In this article, we look at inter-organisational trust and inter-organisational cooperation as the key social aspects in industrial symbiosis. We present the outline of a possible interdisciplinary research programme on industrial symbiosis. This programme would not only productively combine hard and soft sciences, but would also have important practical implications (i.e. development of tools for embedding industrial symbiosis in natural, technical, and socio-cultural setting). It is a research programme dealing with dynamic complex networks, crossing the boundaries and being shaped by, the technical, natural, computer, and social systems. In this research programme, industrial symbiosis would be defined as sharing services, utility, or by-product resources between industries in order to increase value and reduce financial and environmental costs.

Keywords: industrial symbiosis, social aspects, inter-organisational trust, inter-organisational cooperation, social theories

Introduction

Industrial symbiosis represents a relatively new interdisciplinary research area. In terms of a research area, it belongs to ecologic, economic, and sociologic areas. Some studies (Chertow 2007, pp. 12; Howard-Grenville et al. 2008, pp. 158; Gingrich 2012, pp. 44) deal with industrial symbiosis as part of industrial ecology, whereas others see it as a sort of approach to industrial ecology. Industrial symbiosis also appears as a synonym for industrial ecology (Phillips et al. 2005, pp. 242), a subset of industrial ecology (Chertow et al. 2005, pp. 6535), an industrial symbiotic network (Jacobsen et al. 2005, pp. 313), and sometimes even appears inside industrial ecosystems (Rui et al. 2010, pp. 1). Research in industrial symbiosis mainly focuses on industrial ecology and industrial ecosystems and represents positive aspects in ecologic, economic, and sociologic views. In light of ecology, the positive environmental effects appear mainly in the form of less waste/by-product disposal and in the saving of the limited water supply. Positive economic effects are demonstrated in lower waste/by-product disposal costs, competitive advantage towards other economic actors, and in better reputation of the company/economic actor. In the sociological/social sphere, the positive aspects for nearby and more remote societies mostly appear in the awareness that economic actors care for the environment and try not to burden it with unnecessary environmentally unfriendly activities, or if these activities are necessary, to limit them to the minimum.

Social aspects have a key role in the exchange, which is the elementary activity in industrial symbiosis. Two of the most important social aspects in sociological terms are cooperation, involving communication, and trust. Namely, the effectiveness of industrial symbiosis depends strongly on trust and reciprocity (i.e. cooperation between economic actors involved in industrial symbiosis) (Domenech et al. 2010, pp. 281–282). Some authors find that despite the well-know relationship between cooperation, trust, and communication, their roles in industrial symbiosis are not clearly defined (Chertow 2000; Gibbs 2003; Jacobsen et al. 2005 in Ashton 2008, pp. 35). For the scientist, studying social aspects of industrial symbiosis presents many challenges on the one hand and many dilemmas on the other hand due to lack of a specific or explicit methodology for measuring social aspects. The main dilemma in researching these aspects is the selection of research paradigm – whether to select quantitative or qualitative paradigm. Later, we mention the dilemma on whether it is possible to select both research paradigms and whether triangulation methods can be used. In turn, these dilemmas involve further questions on appropriate sampling, selection of data collection methods, data processing methods, and the potential pitfalls of research and how to avoid them. These dilemmas and questions represent the key motivation for our research.

Defining Industrial Symbiosis: Theory and Concepts

Industrial symbiosis is depicted in light of the opportunities for exploiting ecologic and economic synergies, while taking place in the social milieu. Its theory and concepts are described in scientific/research and applicative terms, where companies as social or economic actors offer other potential economic actors paradigmatic solutions in industrial symbiosis in the form of consulting and preparation of programs/projects. Chertow (2007, pp. 12) finds 1989 an inspirational year in industry and the environment, with the beginning of many international studies on the subject of industrial symbiosis. It all started with three internationally acclaimed case studies. The Kalundborg study remains one of the paradigmatic models of the industrial symbiosis of networks (Lowe et al. 1995, pp. 49). In Kalundborg, Denmark, companies and other economic actors in connection with their municipalities developed a complex network of waste material exchanges to be processed and reused as raw material in industrial processes. Eleven companies were involved in the project, which encompassed 13 projects (Van Berkel et al. 2008, pp. 1272). The stability of the network was supported by a so-called positioning process, which represented the foundation for networking and connected the companies in various ways (Domenech et al. 2010, pp. 284). The next case study, as well as an ongoing programme, is the acclaimed National Industrial Symbiosis Programme (NISP), inspired by Peter Laybourn in 1999 and founded in 2003 after the realisation of three pilot projects in Scotland. Since 2003, the NISP has been active in Great Britain as the first voluntary programme looking for possibilities of reusing industrial waste and other input materials in other industrial processes through the paradigm[1] of industrial symbiosis. The NISP includes micro, small, and medium enterprises, as well as multinational corporations from different industries (International Synergies, 2014). In Sagunto, Spain, companies in the steel industry formed a complex network for reusing waste by-products of steel production as raw materials for the metal products industry (Domenech et al. 2010, pp. 286). This project was motivated by economic and ecologic goals, whereas the achievement of these goals depended on geographical location of the companies involved—companies were based in the same city or region. Chertow (2000, pp. 314) states that the key to industrial symbiosis is the cooperation and synergy possibilities enabled by geographical proximity of partners. On the local level, industrial symbiosis in the context of reusing by-products and waste appears in different industries, mainly in the feedstuff and chemical industry (fertilizer production), energy sector (biogas production), and increasingly in agriculture.

Industrial symbiosis can be seen as a relationship between two or more economic actors exchanging materials, by-products, and waste products, where each actor adds to the welfare of other actors and the society in general, while following their own benefits. Chertow (2000, pp. 314) defines industrial symbiosis as a collective approach, the purpose of which is to gain a competitive advantage by

---

1 Etymologically, the term paradigm can be understood as a series of interconnected assumptions on social phenomena which provide a philosophical and conceptual framework for research (The American Heritage Dictionary of the English Language, 2000).

providing for the physical exchange of materials, energy, water, and by-products between economic actors (companies) in different industries. However, industrial symbiosis can only be achieved between economic actors through cooperation and synergy, which depends on the geographical proximity of these partners (Chertow 2000, pp. 314). In practice, it appears as a collective multi-industrial approach to increasing economic and ecologic effectiveness by reusing waste products and/or by-products as raw materials (Costa et al. 2010, pp. 984). The following is an example of industrial symbiosis between two economic actors. One economic actor sells the waste products, by-products, or other materials, produced during their primary activity, to the other economic actor, who processes these materials and uses them as raw materials or input materials for their primary activity. There are three flows connecting the two partners: material, information, and financial flow. Material flow begins with the partner selling the material resources and ends with the partner buying them. Information flow can circulate between both partners, whereas the financial flow usually runs in the opposite direction to material flow. In practice, financial flow can be excluded due to mutual material exchange between the partners. This mostly applies to agriculture, where, for example, animal excrement fertilizer can be exchanged for crops. Economic actor can process the material resources they wish to reuse as input material on their own or can pay another economic actor for the processing. Chertow (2007, pp. 13) has defined a minimum criterion of industrial symbiosis, which demands the involvement of at least three economic actors in the exchange of at least two material resources.

In accordance with this criterion, the already mentioned example between two economic actors represents a "2–1 heuristic", where the same material resource is being exchanged between two economic actors. Chertow (2007, pp. 13) states three possible sources of exchange in industrial symbiosis. One option is to exchange by-products for them to be reused, the second option is utility sharing, and the third option is the collective provision of services. Namely, the subject of exchange depends on the type of companies involved (industrial, non-industrial, agriculture), therefore providing varied and numerous definitions of industrial symbiosis. Despite the differences in its definition, some characteristics of industrial symbiosis remain uniform: economic actors come from different industries, more than one material resource is being exchanged, and cooperation is applied in the form of networking (Chertow 2007, pp. 12).

Social Aspects of Industrial Symbiosis: Inter-organisational Trust and Inter-organisational Cooperation

Definitions, concepts, and findings on social aspects are adapted to research fields, making it hard to find one general definition, concept, or finding, which could be used in every research study on social aspects. Social aspects, however, are key to the formation and realisation of industrial symbiosis. Kurup (2007, pp. 19) states that social benefits of industrial symbiosis materialise through trust and cooperation between partners or organisations taking part in the symbiosis. Trust between economic actors decreases transaction costs, risk, and the uncertainty of the exchange, while being extremely important for the formation of cooperation structures and cooperation itself (Gibbs 2003; Chertow 2007; Baas 2008; Ehrenfeld et al. 1997 in Domenech 2007, pp. 92). Lack of trust within industrial symbiosis complicates and inhibits communication and motivation for cooperation between economic actors (Gibbs 2003; Chertow 2007 in Domenech, 2007, pp. 90). Accordingly, this thesis represents the basis for finding causality between social aspects, whereas we need to understand that there is no universal causality and that no causality can be applied to all instances; neither in industrial symbiosis nor in networking between arbitrary organisations. When both communication and motivation exist between industrial symbiosis' economic actors, trust and cooperation can be seen as the main social aspects for the realisation of its ecologic and economic goals. Communication and motivation as part of industrial symbiosis can be viewed as social aspects, as factors of influence on social aspects, or even as the conditions for social aspects. In our research, we focus on two social aspects—trust and cooperation—representing key factors of successful and effective exchange in industrial symbiosis. With regards to trust, we focus on inter-organisational trust because industrial symbiosis not only implies trust within one organisation (i.e. organisational trust) but demands trust between several organisations or economic actors. Despite focusing on inter-organisational trust, we accept that interpersonal trust within one organisation can result in organisational trust and, consequently, in inter-organisational trust. In light of cooperation, we focus on inter-organisational cooperation, applying the same principles as for trust, because industrial symbiosis not only involves cooperation within one organisation unit. It rather means cooperation between e.g. several organisations or economic actors.

Inter-organisational Trust as a Social Aspect in Industrial Symbiosis

Research on trust is mainly performed in psychology, sociology, economy, and management, resulting in many different definitions and findings on trust (Paliszkiewicz 2011, pp. 15). Trust has a significant role in every aspect of social life (Paliszkiewicz 2011, pp. 15) and represents a precondition for successful organisational cooperation and effectiveness (Lane 1998, pp. 1). In modern society, which is characterised by increasing complexity and uncertainty of the business world, an undisturbed operation would be much more difficult without interpersonal and inter-organisational trust (Lane 1998, pp. 1). Our research focuses on relevant sociological definitions of trust, where trust represents a method of establishing a relationship and a means for its preservation (Sztompka 1999, pp. 25). Trust is a simplification strategy, allowing us to adapt to the complexity of the social environment we live and work in (Sztompka 1999, pp. 25). In sociology, trust can be studies in interpersonal, inter-organisational, and systemic levels (Lane 1998, pp. 2). Kenning (2008, pp. 463) indicates similar findings, saying that trust exists on personal, organisational, and technological levels, whereas he distinguishes between trust within the organisation and inter-organisational trust. In our research, we focus on inter-organisational trust between organisations or economic actors in industrial symbiosis. This brings economic benefits and simplifies and improves cooperation between organisations, also showing in relational marketing and timely supply (Sydow 1998 pp. 31). It can be defined as a result of subtle and recursive bilateral impact on operation and structures (Sydow 1998, pp. 32). Sydow (1998, pp. 32) believes that inter-organisational trust involves a combination of personal, systemic, and procedural trust. Cummings et al. (1996, pp. 303) state that inter-organisational trust is actually about individuals' expectations regarding the individual-individual relationship or individual-group relationship, namely, that the other individual or group will act in good faith and in accordance with their commitments. Consequently, this implies that the individual or group will be sincere in all their commitments and will not take advantage of the other party, should the opportunity arise (Cummings et al. 1996, pp. 303). Inter-organisational trust is very important, yet hard to develop and maintain (Sydow 1998, pp. 32). The existence of inter-organisational trust in industrial symbiosis can be subject to several factors or influences and can be influenced by the level of trust within an organisation or economic actor, and by the form of their organisational culture. It could also be impacted by the branch of economic actors and the related nature of the work, the number of employees per organisation, and by multiculturalism and its related flexibility in industrial symbiosis. In industrial

symbiosis, inter-organisational trust is linked to inter-organisational cooperation, which can, ideally, result in inter-organisational trust as the foundation for
the exchange of material and immaterial resources. Here, one thing needs to be
stressed. The connection mentioned above applies only to those cases, where
inter-organisational trust is based on business or personal motives by individuals
or groups in specific organisations, whereas these motives could have a positive
or negative impact on everyone involved in industrial symbiosis. The opposite
also applies: inter-organisational cooperation results in inter-organisational trust,
although it is possible to have inter-organisational cooperation and economic
actors working together without having inter-organisational trust.

Inter-organisational Cooperation as a Social Aspect in Industrial Symbiosis

Cooperation can be defined as any adjustment, developed at least in part for the
purpose of reproductive success between social actors (Gardner et al. 2009, pp. 1).
Diekmann et al. (2001, pp. 1) state that, in sociology, cooperation represents one
of the oldest and most respected subjects, initially analysed in light of social order
and solidarity, embodying the common values, norms, and appropriate behaviour.
Industrial symbiosis emphasises connection, community, and cooperation between organisations (Ehrenfeld, 2000 in Hatefipour 2012, pp. 26). Later, we focus
on inter-organisational cooperation between organisations or economic actors
in industrial symbiosis. Young et al. (2006, pp. 109) state that sectors, organisations, and chains without inter-organisational cooperation are unable to develop
or attain competitiveness in the market. Inter-organisational cooperation can be
defined as common coordinated actions between several organisations with the
purpose of reaching common goals (Anderson et al. 1984, pp. 45). Fhionnlaoich
(1999, pp. 1) discusses inter-organisational cooperation in a wider sense. He calls
it a process, during which organisations cooperate, because the benefits they can
achieve with inter-organisational cooperation are hard to achieve by themselves.
Inter-organisational cooperation can be categorised as formal or informal, whereas the formal type is defined within contractual obligations and the informal type
is more flexible and adjustable due to unspecific contractual terms (Fhionnlaoich
1999, pp. 1). Industrial symbiosis requires cooperation between organisations
(i.e. inter-organisational cooperation) (Chertow 2012 in Saloluoma 2014, pp. 30)
because all so-called primary activities of industrial symbiosis are based on it. In
industrial symbiosis, inter-organisational cooperation is closely linked to other
social aspects, predominantly with inter-organisational trust, which is also described in this article. Inter-organisational cooperation represents a foundation

not only for current primary activities of industrial symbiosis, but also for the planning and realisation of operational, tactical, and strategic goals. Long-term inter-organisational cooperation can lead to inter-organisational trust between economic actors, resulting in better communication, control, and adaptation to market changes, formation of common politics for their long-term presence on the market, and implementation of innovative processes in the primary activities of industrial symbiosis.

Outline of a Truly Interdisciplinary Research Programme on Industrial Symbiosis

We can see that industrial symbiosis provides ample opportunities for truly inter-disciplinary research with substantial value-added. In fact, we see it as the outline of the entire international research programme, which would not only combine hard and soft sciences in unprecedented ways, but would also have important practical applications (i.e. the development of tools for embedding industrial symbiosis in natural, technical, and socio-cultural setting). The proposed research programme is an interdisciplinary research project dealing with dynamic complex networks, crossing the boundaries, and being shaped by, the technical, natural, computer, and social systems. This is possible as industrial symbiotic networks are such networks. Industrial symbiosis would, in this case, be defined as shar-ing services, utility, or by-product resources between industries, in order to add value and reduce financial and environmental costs. Research would thus also contribute to studying sustainable industrial ecosystems through interactions and exchanges between industrial flows and their surrounding environment, in which sustainability would not only be understood in ecological or environmental terms. We should emphasise that this field of research is steadily gaining ground, as the intentional development of novel symbiotic networks in accordance with the systemic approaches of industrial ecology and industrial symbiosis has significant potential to decrease dissipation of energy and materials. This leads to the interest of researchers, policy-makers, and businesses, which provides the opportunities.

Any ambitious research programme should be planned as a ground-breaking effort, set to provide the firm interdisciplinary theoretical and methodological foundations for systematic research on dynamic industrial symbiotic networks. The aim should be to develop a theoretical model and adopt novel approaches and analytical techniques that will enable us and other researchers to gain new knowledge of the underlying foundations of industrial symbiosis and its observ-able facts.

Starting such a programme, one should divide the research enterprise into two main phases. The first phase should consist of the critical analysis of the literature and the development of new analytical protocols. One should critically overview current state-of-the-art published research and analytical techniques and explore ways to significantly improve them for future research. Additionally, one should study the underlying structure of existing industrial symbiotic networks, from both quantitative and qualitative point of view; using quantitative objects to theoretically model industrial symbiotic networks and design efficient algorithms for their analysis and visualisation and using qualitative research not only to fill any gaps, but to explore completely new ideas and hints.

In the second phase one should test newly developed analytical techniques on reliable data, if possible from a number of countries. This would allow us to analyse the robustness of analytical techniques for the analysis of networks at different stages of development. Currently, the United Kingdom is the only country in the world with a well-developed and functioning NISP. On the other hand, we have more, but still relatively few countries where industrial symbiotic networks are at the early stages of formation. Finally, there are a number of countries with no systematic efforts and where only spontaneously small-scale networks exist. However, this could change in the future. Such countries can be especially interesting cases from the perspective of research on embedding industrial symbiotic networks in environmental, technical, and socio-cultural settings. Slovenia, for example, is an especially intriguing case. Slovenia produces more than 1.9 million tons of residential and commercial waste annually, which includes more than 4000 tons of hazardous waste, which steadily increases year-by-year. Currently more than 50 % of the waste is landfilled. In the future, relevant efforts will be made, as the country's priorities are set to address the use of waste in accordance with Regulation 2150/2002/EC (The European Parliament and the Council of the European Union on Waste statistics in 2002) with a view to increase the use of clean technologies, set appropriate product design, and recovery and energy use. Thus, it can provide an excellent testing ground for new research methods.

The general objective of such a project would be to use mathematical theory (discrete mathematics, networks, statistics) in connection with other disciplines (sociology, computer sciences, environmental studies) to develop tools to research, analyse, design, optimise, and model (computational modelling) dynamic complex industrial symbiotic networks (i.e. their structure, dynamics, and properties) and to determine how they are shaped by their interaction with technical, natural, computer, and social systems.

The research programme would follow a number of other more specific objectives:

1. Reviewing and consolidating state-of-the-art knowledge and research methods on industrial symbiosis, including knowledge on industrial symbiosis practice, procedures, frameworks, and guidelines, as well as a review of scientifically verified technologies enabling these processes (decision support technology, ontology engineering, social networking, and user centric technologies).
2. Developing sound system-theoretic and theoretical foundation and research methods for the analysis of complex industrial symbiotic networks, by exploring applicability of mathematical network analysis.
3. Developing knowledge models and knowledge methodologies capable of capturing and processing knowledge and information in industrial symbiotic networks. In this, it should especially focus on the development of computer ontologies, which are required to manage and facilitate industrial symbiosis implementation.
4. Investigating the mutually structuring communication of industrial symbiotic networks with the technical, natural, computer, and social systems.

Last but not least, a comprehensive research programme should especially emphasise the role of social systems on dynamics of industrial symbiotic networks (i.e. how networks of relevant actors (social networks), institutions (multi-level environmental and industrial policy and practice), and cognitive frames (public perception) influence their structuration).

Another relevant question is that of the key research networks that such a research programme would employ. Industrial symbiosis networks are complex networks that cross the boundaries of and interact with the technical, natural, computer, and social systems. Hence, to reach the project objectives, it would have to first apply research methods that can help us cross the disciplinary boundaries and analyse the structure of the relationship between significant varieties of entities and, secondly, use disciplinary specific methods when necessary to reach specific objectives. We can hypothesise that the main research methods will be the following.

First, the main research method will be mathematical network analysis. It will enable researchers to cross disciplinary boundaries and include the networks that come from technical, natural, computer, and social systems. To employ them meaningfully, we should, as argued above, acquire data for the network analysis through the selection of waste streams in a number of countries at different stages of development. Tools for network analysis and visualisation and combinatorial optimisation would be especially important here.

Secondly, one should adopt the fuzzy-set analysis – a type of qualitative comparative research method – to explore the impact of social networks, institutions, and cognitive frames on industrial symbiotic networks. The key research question here should be to determine whether these social forces are necessary and sufficient conditions for the successful establishment of the industrial symbiotic network, as well as the possible variety and stability of these causal conditions.

Finally, to collect the less systematic and qualitative data, researchers could use various supporting methods, such as secondary documentary analysis, (online) surveys, focus groups, and structured interviews.

What would be the originality of such research programme? The proposed research programme falls in the framework of strongly interdisciplinary research and is dealing with the topic, which is just coming to the attention of researchers, as well as practitioners, from a variety of scientific fields and industries. As such, the project has the potential to go significantly beyond the current state-of-the-art in the newly developing field of industrial symbiosis research. The main contributions to the science will be:

1. We will make significant headway in a recently established field of research – which is in itself is less than quarter of century old – in which it would develop strong theoretical foundations as well as adopt a rich variety of research methods.
2. In this research programme, interdisciplinary will not only be a buzzword. Researchers will use mathematical network analysis to integrate industrial, natural, computer, and social systems perspectives in one analysis, hypothesising that industrial symbiotic networks are shaped by the specific topographies of these four systems.
3. Most importantly, industrial symbiosis as a field is currently mainly managed by and consists of networks mediated by trained practitioners. Science still has to make a contribution if this field is to be developed. Symbiotic links admittedly constitute a complex task that involves background knowledge from a variety of disciplines. There is a need to significantly upgrade industrial symbiosis towards a systematic and powerful field of exploration, which is attractive to the scientific community, as well as to the public.

Conclusion: Towards Practical Considerations

The point of industrial symbiotic networks is to build collective flows among a plethora of actors leveraging the exchange of resources to benefit both their production efficiency and its environment by reducing carbon footprints, minimising landfill waste, and to save non-renewable and fragile resources; this is what any such programme should take into account. Thus, industrial symbiotic networks exist to exchange resources (i.e. by-products, assets, and services) through closed and sustainable lifecycles and supporting sustainable operation. Local coopera-tion in industrial symbiosis reduces raw material use and waste disposal, while material and energy flows extending outside boundaries of networks increase the system's entropy and cause considerable environmental impacts. The low cost of waste also brings tangible economic benefits to members of industrial symbiotic networks. These networks emerged as a collective, multi-industrial approach to improve economic and environmental performance through the use of wastes and by-products as substitutes for raw materials.

However, in spite of these promising benefits, most environments industrial symbiotic networks are poorly developed, the logics of their operations have not been systematically researched and, as a consequence, both policy makers and industrial participants are unaware of the benefits that the establishment of such networks provides. So, we will conclude with few practical considerations.

Firstly, it is recognised that the key to success of the entire project is in embed-ding the industrial symbiosis in the region. Dealing with industrial symbiosis is relevant not only from a scientific perspective, but also from a policy perspective. It is very important that in any attempts at creating industrial symbiotic networks, business and relevant regional and national policy-makers are involved in project implementation. The process of their involvement in the project has to be practi-cal, evidence-based, and tailored to the needs of businesses and policy-makers.

Secondly, this implies that one of the key challenges is to develop tools for creating an integrative adaptive industrial symbiosis awareness and initiative to produce a viral effect of industrial symbiosis in the business community. This model should take into account the knowledge developed and assimilated in the aforementioned research programme and also identify social, political, and other relevant intangible bottlenecks limiting the implementation of industrial symbiosis, such as the lack of cooperation and trust between businesses and poli-cymakers, legal constraints on environmental results, absence of awareness of industrial symbiosis and industrial process and otherwise related environmental consequences, lack of resources, etc. An adaptive implementation model should be based on 'what works' or evidence-based policy. It should include at least raising awareness and initiative, assessment of the situation and creating vision, building consequences and commitment, improvement of policies and strategy, and implementation and continuous evaluation of improvement. It is important to develop a unified industrial symbiosis implementation model that will provide a generic framework, but is sufficiently adaptive to take cognisance of context-specific nuances and needs. This model will become an integral element of the policy process on which it will be possible to base the development and implementation of industrial symbiosis.

Finally, the purpose of this research and policy-making effort should be to maximise the utilisation of existing resources. To that we may add that information technology (IT) can play and will play an increasingly important role in this process. Efforts at embedding the new industrial symbiosis paradigm must include an application of the relevant IT resources. Special consideration has to be made to ensure that they incorporate products of the future research programme. Having said that, we should add that IT resources do not only include industrial symbiosis-specific services that will be developed – some are already being developed – but also utilise a variety of wikis, online workbooks, and existing social networks. Hence, any policy-oriented research will also include an analysis of the applicability of research into the popularity of social networks, including Facebook, LinkedIn, Twitter, etc.

References

Anderson, James C. / Narus, James A.: "A model of the distributor's perspective of distributor manufacturer working relationship". Journal of Marketing 48(4) 1986, pp. 45.

Ashton, Weslynne: "Understanding the Organization of Industrial Ecosystems: A Social Network Approach". Journal of Industrial Ecology 12(1) 2008, pp. 35.

Chertow, Marian R.: "Uncovering Industrial Symbiosis". Journal of Industrial Ecology 11(1) 2007, pp. 12–13.

Chertow, Marian R.: "Industrial Symbiosis: Literature and Taxonomy". Annual Review of Energy & the Environment 25(1) 2000, pp. 314.

Chertow, Marian / Lombardi, Rachel D.: "Quantifying economic and environmental benefits of co-located firms". Environmental Science and Technology 39(17) 2005, pp. 6535.

Costa, Ines / Ferrao, Paulo: "A case study of industrial symbiosis development using a middle-out approach". Journal of Cleaner Production 18(2010) 2010, pp. 1.

Cummings, Larry L. / Bromiley, Philip: "The organizational trust inventory (OTI): development and validation". In: Kramer, Roderick Moreland / Tayler, Tom R. (eds.): *Trust in Organizations: Frontiers of Theory and Research*. Thousand Oaks Sage: USA 1996, pp. 303.

Diekmann, Andreas / Lindenberg, Siegwart: "Sociological aspects of cooperation". In: Smelser, Neil J. / Baltes, Paul B. (eds.): *International Encyclopedia of the Social and Behavioral Sciences*. Elsevier: Amsterdam 2001, pp. 1.

Domenech Aparisi, Teresa A.: *Social Aspects of Industrial Symbiosis Networks*. (Bartlett School of Graduate Studies University College London). (doctoral thesis) London 2007, pp. 90–92.

Domenech, Teresa A. / Davies, Michael: "The Role of Embeddedness in Industrial Symbiosis Networks: Phases in the Evolution of Industrial Symbiosis Networks". Business Strategy and the Environment 20(2010), pp. 281–286.

Fhionnlaoich, Cormac Mac: "Interorganizational Cooperation: Towards a Synthesis of Theoretical Perspectives". In: McLoughlin, Damien / Horan, Conor. (eds.): *Proceedings of the 15th Annual IMP Conference*. University College: Dublin 1999, pp. 1.

Gardner, Andry / Griffin, Asleigh S. / West, Stuart A.: "Theory of Cooperation". Encyclopedia of life sciences. John Wiley & Sons 2009, pp. 1.

Gingrich, Caleb: "Industrial Symbiosis: Current understanding and needed ecology and economics influences". Policy engagement: Centre for Engineering and Public Policy, 2012, pp. 44.

Hatefipour, Saeid: "Facilitation of Industrial Symbiosis Development in a Swedish Region". Linköping University, Environmental Technology and Management: Department of Management and Engineering 2012, pp. 26.

Howard-Grenville, Jennifer / Paquin, Raymond: "Organizational Dynamics in Industrial Ecosystems: Insights from Organizational Theory". In: Ruth, Matthias I. / Davidsdottir, Brynhildur (eds.): *Dynamics of Industrial Ecosystems*. Edward Elgar: UK, 1(2008) 2008, pp. 158.

International Synergies, Industrial ecology solutions: *National Industrial Programme (NISP Network)* retrieved 12.11.2014, from http://www.international-synergies.com/projects/national-industrial-symbiosis-programme-nisp.

Jacobsen, Noel Brings / Anderberg, Stefan: "Understanding the evolution of industrial symbiotic networks: the case of Kalundborg". In: Van den Bergh, Jan / Janssen Marijn (eds.): *Economics of Industrial Ecology: Materials, Structural Change, and Spatial Scales*. MIT Press: MA 2005, pp. 313.

Kenning, Peter: "The influence of general trust and specific trust on buying behaviour". International Journal of Retail & Distribution Management 36(6) 2008, pp. 463.

Kurup, Biji R.: *Methodology for Capturing Environmental, Social and Economic Implications of Industrial Symbiosis in Heavy Industrial Areas*. (University of Technology, Curtin University, Division of Science and Engineering). (doctoral thesis) Perth Australia 2007, pp. 19.

Lane, Chris: "Introduction: Theories and issues in the study of trust". In: Lane, Chris / Bachman, Reinhard (eds.): *Trust within and between organizations, conceptual issues and empirical application.* Oxford University Press: Oxford 1998, pp. 1–2.

Lowe, Ernest A. / Ewans, Lavrence K.: "Industrial ecology and industrial ecosystems". Journal of Cleaner Production 3(1/2) 1995, pp. 49.

Paliszkiewicz, Joanna: "Inter-Organizational Trust: Conceptualization and Measurement". International Journal of Performance Measurement 1(2011) 2011, pp. 15.

Phillips, Paul S. / Barnes, Richard / Bates, Margaret P. / Coskeran, Thomas: "A critical appraisal of an UK county waste minimisation programme: The requirement for regional facilitated development of industrial symbiosis/ecology". Resources, Conservation and Recycling Elsevier: Melbourne, 46(2005) 2005, pp. 242.

Rui, Jiavi / Heijungs, Reinout: "Industrial ecosystems as a social network". Knowledge Collaboration & Learning for Sustainable Innovation ERSCP-EMSU conference, Delft, The Netherlands 2010, pp. 1.

Saloluoma, Mikki: *Developing of supportive context for eco-industrial networks in Finland.* (Lappeenranta University of Technology: Department of Industrial Engineering and Management). (master´s thesis) Finland 2014, pp. 30.

Sydow, Jörg: "Understanding the Constitution of Interorganizational Trust". In: Lane, Christel V. / Bachamann, Reinhard (eds.): *Trust Within and Between Organizations: Conceptual Issues and Empirical Applications.* Oxford University Press: Oxford 1998, pp. 31–32.

Sztompka, Piotr: *Trust: A sociological Theory.* Cambridge University Press: Cambridge 1999, pp. 25.

The American Heritage Dictionary of the English Language: Paradigm, retrieved 17.11.2014, fromhttps://www.ahdictionary.com/word/search.html?q=paradigm&submit.x=0&submit.y=0

Van Berkel, Rene / Fujita, Tsuyoshi / Hashimoto, Shizuka / Fujii, Minoru: "Quantitative Assessment of Urban and Industrial Symbiosis in Kawasaki, Japan". Environmental Science & Technology 43(5) 2009, pp. 1272.

Young, Louise / Wilkinson, Ian F.: "The role of trust and cooperation in marketing channels. European Journal of Marketing 23(2) 2006, pp. 109.

Petar Juric
Department of Information Technology Services
Primorsko-goranska County
Adamiceva 10, 51000 Rijeka, Croatia
petar.juric@pgz.hr

Maja Matetic, Marija Brkic
Department of Informatics
University of Rijeka
Radmile Matejcic 2, 51000 Rijeka, Croatia
{majam, mbrkic}@inf.uniri.hr

Game-based Learning and Social Media API in Higher Education

Abstract: In the forthcoming years, e-learning systems will be increasingly accessed by mobile computer devices, such as tablets and smartphones. Developing computer games for this platform will be a growing share of software development and of new learning methods. This paper provides an analysis of motivational elements of computer games. An overview of the newest technologies for developing computer games in a mobile web environment and of the current research on using computer games for learning programming and for foreign language learning is given. A model on which our future work will be based on is suggested. The model enhances the e-learning system of the University of Rijeka with computer game-based learning methods and the social network communication channel within the course Programming 2.

Keywords: motivation, computer game-based learning, social media, m-learning, e-learning

1 Introduction

Using motivational elements for knowledge acquisition and application is an important research area, which contributes to the learning outcomes. Motivation can be intrinsic (arises from within and it is reflected in the increased activity and the desire to participate evoking a feeling of satisfaction during the activity) and extrinsic (arises from other people's influence (i.e. parents', employers', teachers', etc.)). Specific types of motivation, such as integrative motivation, might be present during foreign language learning in cases where the mastery of a language is a precondition on community integration and communication. In instrumental motivation, the learning desire arises from better job prospects, higher salary, etc., but does not necessarily evoke a feeling of satisfaction (Ozgur / Griffiths 2013).

Current research aims at finding methods for increasing the level of motivation while learning supported by information technology. According to the Gartner's Hype Cycle for Emerging Technologies (Gartner 2013), the term gamification is identified at the peak of inflated expectations.

The concept gamification is originally found in video games. It refers to applying video game principles to everyday life or to applying elements found in games to different products or services in order to encourage their use and add more pleasure when using them (Deterding et al. 2011). Applying the gamification concept primarily affects a person's emotions, or more precisely, a person's intrinsic motivation. Intrinsic motivation has stronger effects and lasts longer. It has a higher engagement level compared to extrinsic motivation (Burke, 2014).

The question that emerges is how computer games can be used in higher education with the goal of increasing students' motivation for new content acquisition.

2 Motivational elements in computer games

E-learning systems play an important role in present higher education systems; however, lack of motivation for using those systems stands out as a main drawback. Newer generations of students (born in the last quarter of the twentieth century) exhibit different prevailing cognitive styles, which originate from computer games (Prensky 2007).

Motivational elements found in computer games are presented in Table 1. They are ordered by importance for the students. The ordering implies where attention should be primarily focused while processing educational content in computer games. The levels of action attached to these motivational elements may be individual or interpersonal.

Table 1: *Motivational elements of computer games (adapted from Hainey et al. 2011)*

Rank	Reasons	Description	Level
1st	Challenge	an appropriate level of difficulty and challenge, multiple goals for winning, constant feedback and sufficient randomness	individual
2nd	Curiosity	providing sensory stimulation	individual
3rd	Cooperation	assist others	interpersonal

Rank	Reasons	Description	Level
4th	Competition	compare performance	interpersonal
5th	Control	the ability to select choices and observe the consequences	individual
6th	Recognition	a sense of satisfaction when accomplishments are recognised	interpersonal
7th	Fantasy	an appropriate level of immersion by assuming a particular role	individual

The research in this paper will be directed towards available technologies and toward the possibility of learning programming languages and foreign languages through computer games.

3 Open source technologies for programming computer games in a mobile web environment

According to the Recommendations for the Formulation of Educational E-Learning Material of the University of Rijeka (The Committee for Implementation of E-Learning at the University of Rijeka 2009) and the E-learning Development Strategy of the University of Rijeka (University of Rijeka 2011), e-learning is based on the internet (web) technologies. The goals listed in the recommendations and the strategy to which our future research will be directed are the following: 1. training by incorporating new technologies and applying new methods of learning and teaching which enable active knowledge acquisition, and 2. opening possibilities for directing university activity towards new target groups of students by developing distance learning programs (courses).

The technology of internet access and the way of consuming web content is changing due to the increase in the sales of smart mobile devices (tablets and smartphones) compared to desktop PCs and laptops. New devices have a touch screen as a primary input device (instead of a keyboard and a mouse) and different sensors which detect device orientation, direction, and the angle at which the device is held, as well as device location and speed.

Smart mobile device applications can be developed in a native or a web-based environment. A short overview of native application development technologies compared to web-based technologies for the two dominant mobile platforms, Android and iOS, is given in Table 2, along with their pros and cons.

Table 2: Android and iOS native application development compared to web-based development – pros and cons

Platform	Programming language	Development environment	Pros	Cons
Android	Java	Eclipse	Fast, a higher number of APIs for accessing hardware sensors	Platform-specific, changes are available only after reinstallation
iOS	Objective-C Swift	Xcode		

Android and iOS platforms are incompatible regarding the applications developed for each of them. Platform-specific application development means that a game created for Android needs to be developed in another development environment and another programming language for iOS (IBM 2012).

From our point of view, the availability of e-learning content should not be platform-dependent, and it should be extendible and available on as many devices as possible. Since any changes in a web-based application are immediately available to all users, the speed benefit of running native applications, which is important for running games due to the high hardware requirements, falls into the background. Moreover, further fragmentation of mobile devices and technologies is restricted. By developing new APIs for HTML5 platform, the benefits of building native applications will be increasingly reduced (Puder / Tillmann / Moskal 2014).

In our future work, the development of educational games will, therefore, be based on: 1. open-source technologies, 2. platform and plug-in independent web environment, and 3. mobile device adaptation. The platform of choice will be the HTML5 ecosystem with responsive design implemented, which satisfies all of the three requirements. Currently, there is no framework of this type for educational game development. The HTML5 game development frameworks generally include collision detection and physical models of the real world (i.e. gravitation (HTML5 Game Engines 2014)). Game development is slow, and in most cases, frameworks do not have GUI, while only a few have a level editor.

The familiarity with technology required of teachers needs to be reduced to the lowest level for the wider acceptability of content processing through educational computer games. We suggest building predefined game types in which teachers could add textual and visual elements to be displayed on the learning objects in the game simply via an administrator interface (Figure 1). This module will form a central part related to learning new content, along with the integrated social media API, to which data mining systems will connect with the task of optimising the learning process.

Figure 1: The process of creating educational computer games based on predefined types

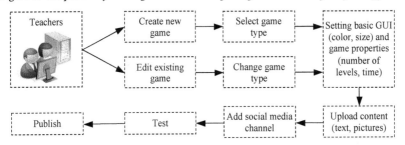

4 Learning through computer games in higher education

Learning through computer games offers almost unlimited application possibilities in which real life situations can be mapped to or imaginatively displayed in a virtual world (De Freitas et al. 2009).

The work will focus on increasing motivation and on easier acquisition of more complex concepts within the course Programming 2 at the Department of Informatics of the University of Rijeka (i.e. concepts for which students obtain lower scores: 1. pointers, and 2. Recursion). Similar research on improving learning of object-oriented programming techniques shows positive results (Depradine 2011).

Intersecting points between learning programming and learning other subjects, primarily English as a second language, will be identified. Common types of knowledge developed and acquired while learning programming and foreign languages are syntactic and conceptual knowledge (Sanchez-Lozano 2010). Current research indicates that increasing motivation level has a greater impact on achievement than a predisposition for language acquisition (Daskalovska / Koleva Gudeva / Ivanovska 2012).

A social media API will be implemented as a main textual communication channel between the students themselves and between the students and the teacher. Besides learning through computer games, using social media for thematic communication can also have a motivating effect (Silius et al. 2010).

5 The integrated model proposed and data mining in e-learning systems with computer games

Systems for learning through computer games can be attached to an e-learning system or they can be integral parts of it. The term m-learning or mobile learning is tied to smartphones. These systems can manage structured and unstructured data suitable for data mining (Juric / Matetic / Brkic 2014). Motivation and exam success can be predicted based on interaction with the game (Illanas Vila et al. 2013). Structured data refers to location, speed, number of level attempts, time necessary to get to the next level, number of replays of the level after completing it, number of hints used, etc. Unstructured data originate in using integrated social media API either for communication or for seeking help from other students or the teacher.

The University of Rijeka uses MudRi learning management system (LMS), which is based on the open-source Moodle system.

The integrated model of a computer game based learning system (GBLS) with social network communication between users is given in Figure 2. The data and text mining systems are attached to the model.

Figure 2: The integrated model of learning through computer games with social network communication channel and data mining within an e-learning system

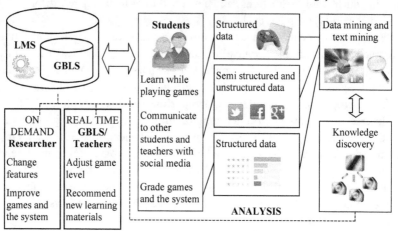

Data mining enables teachers to adapt game contents and game level to the level of students' prior knowledge and to recommend additional learning activities. It enables experts to analyse data in order to improve games by optimization (i.e. with the goal to achieve a faster and more thorough approach to learning). Other data mining systems in the field that do not require expert knowledge might also be attached to the model (Jugo / Kovacic / Slavuj 2013).

For evaluating satisfaction with the learning approach, a short questionnaire at certain points through the game or at the end of the game will be used, as well as game grading and social media API communication.

6 Conclusion

Using computer games in education triggers intrinsic motivational elements. Since the approach is fun, the time students spend learning, as well as the level of knowledge acquisition, is increased. The reason why it is not widely represented lies in programming requirements and unfamiliarity with web technologies and technologies for game development. Besides complexity, another drawback is the time needed for the development. Since these elements have so far been developed in companies or by expert scientific teams, the proposed model, which would enable developing games based on predefined types, provides grounds for the simple and fast development of certain types of educational computer games.

By applying data mining techniques in these systems, the teachers could track the level of engagement and success prior to oral or written examinations, as well as recommend materials or activities for remedial or additional classes.

7 Acknowledgement

This research has been supported under the Grant No. 13.13.1.3.03 of the University of Rijeka.

8 References

Burke, Brian: *Gamify: How Gamification Motivates People to Do Extraordinary Things*. Biblimotion. Brookline, MA, USA, 2014.

Daskalovska, Nina / Koleva Gudeva, Liljana / Ivanovska, Biljana: "Learner motivation and Interest". *Procedia – Social and Behavioral Sciences* 46, 2012, pp. 1187–1191.

De Freitas, Sara et al.: "Developing an Evaluation Methodology for Immersive Learning Experiences in a Virtual World". In: Rebolledo-Mendez, Genaro et al. (eds.): *Proceedings of the Games and Virtual Worlds for Serious Applications VS-GAMES 2009*. IEEE, 2009, pp. 43–50.

Depradine, Colin A.: "Using Gaming to Improve Advanced Programming Skills". *Caribbean Teaching Scholar* 1(2), 2011, pp. 93–113.

Deterding, Sebastian et al.: "From Game Design Elements to Gamefulness: Defining Gamification". In: Lugmayr, Artur et al. (eds.): *Proceedings of the 15th International Academic MindTrek Conference: Envisioning Future Media Environments*, ACM, 2011, pp. 9–15.

Gartner: *Gartner's 2013 Hype Cycle for Emerging Technologies Maps Out Evolving Relationship Between Humans and Machines*, retrieved 21.6.2014, from http://www.gartner.com/newsroom/id/2575515.

Hainey, Tom et al.: "The differences in motivations of online game players and offline game players: A combined analysis of three studies at higher education level". *Computers & Education* 57(4), 2011, pp. 2197–2211.

HTML5 Game Engines: *Which HTML5 Game Engine is right for you?*, retrieved 27.6.2014, from http://html5gameengine.com.

IBM: *Native, web or hybrid mobile-app development*, retrieved 4.7.2014, from http://public.dhe.ibm.com/software/in/events/softwareuniverse/resources/Native_web_or_hybrid_mobile-app_development.pdf.

Illanas Vila, Ana et al.: "Predicting student performance in foreign languages with a serious game". In: Gomez Chova, Luis et al. (eds.): *Proceedings of 7th International Technology, Education and Development Conference INTED2013*. IATED, 2013, pp. 52–59.

Jugo, Igor / Kovacic, Bozidar / Slavuj, Vanja: "A proposal for a web based educational data mining and visualization system". In: Levnajic, Zoran (ed.): *Proceedings of the 5th International Conference on Information Technologies and Information Society ITIS 2013*, FIS, 2013, pp. 59–64.

Juric, Petar / Matetic, Maja / Brkic, Marija: "Data Mining of Computer Game Assisted e/m-learning Systems in Higher Education". In: Biljanovic, Petar et al. (eds.): *Proceedings of the 37th International Convention MIPRO*, IEEE, 2014, pp. 750–755.

Ozgur, Burcu / Griffiths, Carol: "Second language motivation". *Procedia – Social and Behavioral Sciences* 70, 2013, pp. 1109–1114.

Prensky, Marc: *Digital game-based learning*, Paragon House, St. Paul, MN, USA, 2007.

Puder, Arno / Tillmann, Nikolai / Moskal, Michal: "Exposing Native Device APIs to Web Apps". In: *First International Conference on Mobile Software Engineering and Systems MOBILESoft 2014*, ACM, 2014, pp. 18–26.

Sanchez-Lozano, Juan Carlos: "Exploratory Digital Games for Advanced Skills: Theory and Application". In: Zemliansky, Pavel / Wilcox, Diane (eds.): *Design and Implementation of Educational Games: Theoretical and Practical Perspectives*. IGI Global, 2010, pp. 92–107.

Silius, Kirsi et al.: "Students' Motivations for Social Media Enhanced Studying and Learning". *Knowledge Management & E-Learning: An International Journal* 2(1), 2010, pp. 51–67.

The Committee for Implementation of E-Learning at the University of Rijeka: *Preporuke za izradu obrazovnih materijala za e-ucenje* [Recommendations for the Formulation of Educational E-Learning Material], retrieved 22.6.2014, from http://www.uniri.hr/files/vijesti/Preporuke_e-ucenje_2009_UNIRI. pdf.

University of Rijeka: *Strategija razvoja e-ucenja na Sveucilistu u Rijeci 2011–2015* [E-learning Development Strategy 2011–2015], retrieved 22.6.2014, from http://www.uniri.hr/files/staticki_dio/propisi_i_dokumenti/Strategija_ e-ucenje_ 2011–2015.pdf.

Andrej Kovačič, Ph.D.
Faculty of Media
Leskoškova 9D, 1000 Ljubljana, Slovenia
{andrej.kovacic@ceos.si}
Nevenka Podgornik, Ph.D.
School of Advanced Social Studies in Nova Gorica
Gregorčičeva 19, 5000 Nova Gorica, Slovenia
{nevenka.podgornik@fuds.si}

Negatively Biased Media in Slovenia

Abstract: The article focuses on the question of how much negativity is present in Slovenian Media and why. As this important issue seems to be neglected in academic writing, we have conducted several studies analysing RSS headlines and short abstracts on the internet. Despite many methodological obstacles when analysing news on the internet, our research has enabled us to discuss the negativity of news reporting in a wider context. The main objective of this article is to answer an important research question: to what extent does domestic and foreign ("imported") news from different media sources (television, newspaper and internet) evoke negative feelings. As Slovenia is a small country, the lack of domestic negative shocking news is filled with foreign news, where the majority of them evoke negative feelings. Such disproportion consequently portrays a more negative image on the world outside of Slovenia. This could lead to a misconception of foreigners and their culture and can seriously influence migration characteristics of Slovene people, which are one of the lowest in the European Union.

Keywords: negative news reporting, RSS, foreign, home

Introduction and literature preview

A discussion about the consequences of negative news reporting has been abundant. Johnston and Davey (1997) tested three groups with 14-minute television (TV) news bulletins that were edited to display either positive, neutral, or negative material. Negative bulletin participants showed "increases in both anxious and sad mood", and a "significant increase in the tendency to catastrophise a personal worry". Similarly, Johnston and Davey suggest that negative TV news programs can increase personal concerns that are not specifically relevant to the content of the program. What is of additional importance is that increases in negative mood as a result of viewing a negative news bulletin were also associated with increases in the catastrophising of personal worries. Johnston and Davey also report that negative programs are likely to have a negative effect on mood by increasing an individual's personal worries and anxieties. In fact, as Horn (2007) explains, public discontent with the media in general largely results from exaggerated claims based on erroneous reasoning along these lines. Extreme consequences of negative news reporting can be found in Sudak H. and Sudak D. (2005), as they provided examples of how specific media reporting of suicide news has an impact on further suicides. Apollonio and Malone (2009) analysed negative political reporting and political advertising, another context of negative news reporting, and concluded that by increasing awareness to potential problems, media is actually encouraging people to reconsider their established opinions. Browne and Hamilton – Giachritsis (2005) similarly analysed violence in connection to the violence presented in media. They believe that long-term outcomes for children viewing media violence are controversial because of the methodological difficulties.

The relationship between mood influenced by media and long-term health risk is pointed out in a thirty-five-year longitudinal study by Peterson, Seligman and Vaillant (1988). This study indicates that the pessimistic style predicted poor health at ages 45 through 60, even when physical and mental health at age 25 were controlled. Thus, pessimism in early adulthood appears to be a risk factor for poor health in middle and late adulthood.

Slovenian media research is limited. A decade old research study was done by Petrovec (2003), who claims media reporting is focused on the most spectacular stories, mainly presenting violence in volume, and is not representing the reality. In addition, Petrovec claims that negative news reporting is contributing to the idea of bringing fear in Slovenian society. In his research, almost 80 % of Slovenian people are convinced that Slovenian media is too full of programs containing violence and thus urge for a sort of guardian of public interest in media reporting. In POP TV, the share of news containing violence was 40.3 % and in TV SLO 1 program, 18.2 % (Petrovec 2003). Similarly in a tabloid, Slovenske noticed that the share of news containing violence was 26.9 %, whereas the share of words in headlines implying violence was over 80 %.

Cummings (2007) defines engagement as client specific emotional response. As such, Heath (2007) claims that not achieving attention or engagement means you are in serious danger of having an unsuccessful message. "Bad" news attracts more attention than "good" news. The incentive to buy (purchase newspapers or spare television time) is so closely related to visual attention. Visual attention can be explained (Ketelaar et al. 2008) as the allocation or concentration of cognitive energy to process one part of the visual field at the expense of other parts. Young (2010) refers to this objective as "stopping power". For advertising, he states that in order to be effective "print advertisement must get noticed and attract a reader".

Research hypotheses, research design, method, and sampling

We set the research question as to whether or not there is a difference in the Slovenian and foreign news reporting tone. We hypothesise that a small country that does not have enough shocking domestic news tends to use foreign negative news to fill in the gap. This can lead to many misperceptions of the outside world and can influence perception and migration preferences.

We used a random sample for each media with each record (news) within the group having exactly the same probability of being selected without repetition. The number of evaluated articles (sample number) can be seen in Table 1.

Table 1: Sample numbers for different media

Language group	Sample number	Domestic news topic	Foreign news topic
TV	651	305	346
Newspaper	977	569	408
Free newspaper	648	366	282
Internet	330	200	120
Total	2606	1166	1440

Source: own research, 2013 and 2014

This selection procedure assured the normal distribution as well as the sufficient representation of each of the selected media. The selection procedure was based on the readership ratings of eight selected sources. All media is national media (no local media is represented as Slovenia is a small country). In the analysis, the independent variable was the evaluation of the story based on a one item nine-point Likert scale, a pleasure dimension of Self-Assessment Manikin (SAM) (Appendix 1) – "How did reading this news make you feel?" SAM is a visual non-verbal scale with a graphic character arrayed along a continuous nine-point scale with Cronbach alpha = 0.82 and 0.98 (see also Backs et al. 2005; Morris et al. 2002; Peterson et al. 1988; Poels an Dewitte (2006); and Sudak and Sudak (2005)).

Figure 1: How did reading this news make you feel?

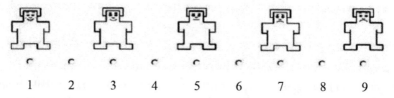

1   2   3   4   5   6   7   8   9

Source: adapted from Moris (1995), (Moris et al. 2002)

Results of the evaluation were calculated to range from -4 (extremely negative evoked feelings) through 0 (neutral) to 4 (extremely positive evoked feelings). Although the Likert scale is in its essence ordinal in this example, SAM can approximate an interval-level used for measurement ranging from -4 to +4. A total of 2606 news sources were evaluated, with each news source being evaluated twice to test the reliability of the evaluators.

Inter-rater reliability calculations

To improve the reliability of this research, each news headline and short summary (up to 250 characters) was evaluated twice by two evaluators. The evaluators were independent and were financially compensated for the task. Their judgments were not influenced by any of the authors or support team and none of the authors of this article were a part of the evaluation team (consisted of four evaluators in this study). Evaluators were native speakers living in Slovenia. Following the evaluation, a Krippendorff's alpha (Krippendorff 2006) on interval variables was used to evaluate the inter-rater reliability.

The basic formula for evaluating the Krippendorff's (2006) alpha is:

$$_{nominal}\alpha = 1 - \frac{D_o}{D_e} = \frac{A_o - A_e}{1 - A_e} = \frac{(n-1)\sum_c o_{cc} - \sum_c n_c(n_c - 1)}{n(n-1) - \sum_c n_c(n_c - 1)}$$

where

$$D_o = \frac{1}{n}\sum_c \sum_k o_{ck\,metric}\delta_{ck}^2$$

and

$$D_e = \frac{1}{n(n-1)}\sum_c \sum_k n_c \cdot n_{k\,metric}\delta_{ck}^2$$

Krippendorff's alpha is exclusively rooted in the data generated by all observers and defines the two reliability scale points as 1.00 for perfect reliability and 0.00 for the absence of reliability which are statistically unrelated to the units they describe. The sampling distribution of the means is assumed to be normally distributed as well as the sampling distribution of the scores. Reliability calculations for media tone using Krippendorff's alpha showed alpha = 0.926. Calculated reliability for evaluation on SAM showed alpha = 0.945. We also tested the reliability for the separation of foreign and domestic news and it showed alpha = 0.959. All calculated alphas are to be considered a reliable variable for analysis. Krippendorff (2011) suggests relying on evaluators with variables $\alpha \geq 0.80$, although $\alpha \geq .667$ can suffice for drawing tentative conclusions. For example, alpha 0.90 % means that 90 % of the units tested by evaluators are perfectly reliable while only 10 % are a result of chance. In addition, 55 news articles from different sources were compared using the RSS feed and actual broadcasted/printed material and were thus tested to assure that the RSS feed is indeed the actual news reporting. All 55 articles were identical and thus we can conclude that the RSS feed is indeed a precise reflection of actual media reporting.

Results

A One-Way Analysis of Variance (oneway ANOVA) with Post Hoc Comparisons was used to analyse whether means for Slovenian and foreign news differ. The analysis of the independent variable was the evaluation of the story based on a one item nine-point Likert scale. The results of the evaluation were calculated to range from -4 (extremely negative evoked feelings) through 0 (neutral) to 4 (extremely positive evoked feelings).

The difference between foreign and Slovene news tone can be clearly seen in Figure 2.

Findings:

- There are substantially less articles that are positively portraying foreign news.
- The comparison ratios positive:negative are
  o For Slovenian news – positive:negative= 1:2.37
  o For foreign news – positive:negative= 1:4.19

Figure 2: Difference between foreign and Slovene news reporting in a count of articles

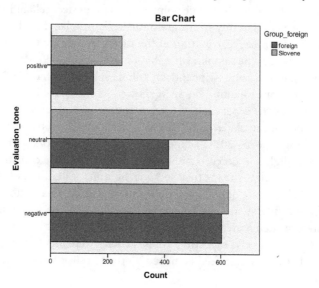

Source: Own research, 2013

The results from the analysis indicate that there is a significant difference be-tween the two groups $F(3,2606) = 9.224$, $p < .001$. The mean values (-0,63) for the evaluations of the Slovenian news differ from evaluations of foreign news (-1,03).

Figure 3 shows us that seven out of eight media sources portray more nega-tive foreign news than Slovene news. The only exception against the suggested hypothesis is Finance (a financial newspaper). Altogether, media portrays on average from 0.1 to 0.75 points on a nine-point likert scale, with a lower evalu-ation for foreign news (lower is more negative). Thus, we can confirm our hy-pothesis that media indeed selects more negative foreign news and less positive foreign news.

Figure 3: Difference between the average foreign and Slovene news reporting evaluation in different media

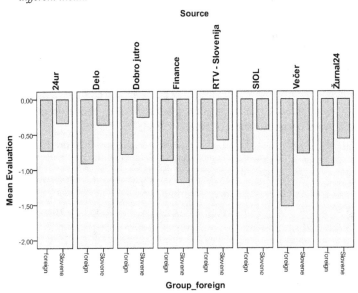

Source: own research, 2013

Limitations and future research

Every research study has some limitations and this one is no exception. We understand that although being chosen on a descending order according to their popularity, the chosen media may not correctly represent the nation's media. As a consequence, we attach the analysis of separate media to allow the reader to separate and make his/her own conclusions. For further research, we would also suggest for the stratification of the media type on a national and international level in addition to popularity. This however exceeded the scope of this research.

We also acknowledge the fact that evaluators could judge news according to the value standards in their own environment. Therefore, a lot of effort has been invested to train the evaluators which is indicated by strong inter-coder reliability results. In addition, the research studies' clear methodology enables other researchers to verify these results and thus contribute to better understanding of the field.

Despite these limitations, it is very important to address this issue in the future. We suggest additional research to focus on the international comparison, where data from different language groups could be compared. In addition, it would be interesting to discuss further implications of negative news reporting in correlation to migrations and health.

Conclusions and implications

So much negative news, especially from abroad, is a broad and systemic problem of media reporting. This practice is stimulated by limited media attention and profitable media outcomes. Even more, by being able to change values and criteria of the people towards extremes of what is acceptable, the tendency is strongly leaning towards even more negative news reporting.

Media is strongly biased towards negative news reporting and as such (ab)uses foreign news to fill in the gap. What happens to the image of foreigners if you portray foreign news that is very negative? This research has proven that Slovenia as a small country imports a large number of foreign news that evokes negative feelings. This results in portraying a more negative image on the world outside of Slovenia and might lead to a misconception of foreigners and their culture. To what extent negative news can influence migration characteristics of Slovene people is a complicated question and needs to be addressed in future research.

Finding a solution to this serious issue of negative news reporting is difficult. On the one hand, regulation can improve the ratio of the negativity in news reporting as also recognised by Horn (2007). On the other hand, media is lobbying strongly for deregulation, promising existing politicians support on elections. Nevertheless, we have to consider the results of this study to provide a strong incentive for future research and legislative action.

References

Apollonio, D. E. / Malone, R. E.: "Turning negative into positive: public health mass media campaigns and negative advertising". Oxford: Health Education Research – Oxford University Press 24 (3), 2008, pp.: 483–495.

Backs, Richard W. / Silva, Sergio P. / Han, Kyunghee: "A Comparison Of Younger And Older Adults' Self-Assessment Manikin Ratings Of Affective Pictures". Experimental Aging Research. 31, 2005 pp.: 421–440.

Browne, Kevin D / Hamilton-Giachritsis, Catherine: "The influence of violent media on children and adolescents: a public-health approach". London: The Lancet (Elsevier Limited). 365 (9460), 2005, pp.: 702–710.

Cummings, Maria N.: "Consumer Engagement Perspectives: A Tool for Ensuring Advertising's Impact?" Rochester: School of Print Media Rochester Institute of Technology. Rochester, New York. 2007

Heath, Robert (2007): "Emotional Persuasion in advertising: A Hierarchy-of-Processing Model". Bath: University of Bath. 2007

Horn, Karen: "A Market Like Any Other: Against the Double Standard in Judging the Media". Oakland: The Independent Review. 12 (1), 2007, pp.: 27–46.

Johnston, Wendy M / Davey, Graham C. L.: The psychological impact of negative TV news bulletins: The catastrophizing of personal worries. London: British Journal of Psychology. 88, 1997, pp.: 85–91.

Ketelaar, Paul E. / Gisbergen, Mamix S. / Van, Bosman, Jan A.M. / Beentjes, Hans: "Attention for Open and Closed Advertisements. Journal of Current Issues in Research in Advertising". 25 (2), 2008, pp.: 30–15.

Kovačič, Andrej: "How do Media Report News in Slovenia? Is there a negative Bias in Communication?" In Viera Žuborova, Diana Kamelia & Uroš Pinterič: Social responsibility in 21st century. Ljubljana: Vega. 2011

Krippendorff, Klaus: "Testing the Reliability of Content Analysis Data: What is Involved and Why". Pennsylvania, ZDA: University of Pennsylvania. 2006, Available at: http://repository.upenn.edu/asc_papers/43 (12.5.2014)

Krippendorff, Klaus: "Computing Krippendorff's Alpha–Reliability". Pennsylvania, ZDA: University of Pennsylvania. 2011, Available at: http://repository.upenn.edu/asc_papers/43 (9.2.2014)

Moris, Jon D.: "SAM: Self Assesement Manikin – an Effecient Cross-cultural Measurement of Emotional Response". Journal of Advertising. 38(1), 1995, pp.: 123–136.

Morris, Jon D. / Woo Chongmoo, Geason / James A. / Kim Jooyoung: "The Power of Affect: Predicting Intention". Journal of Advertising Research., 2002, pp.: 7–17.

Peterson, Christopher / Seligman, Martin E. / Vaillant, George E.: "Pessimistic explanatory style is a risk factor for physical illness: A thirty-five-year longitudinal study". Journal of Personality and Social Psychology. 55 (1), 1988, pp.: 23–27.

Petrovec, Dragan: "Mediji in nasilje:obseg in vpliv nasilja v medijih v Sloveniji". Ljubljana: Mirovni inštitut. 2003

Poels, Karolien / Dewitte, Siegfried: "How to Capture the Heart? Reviewing 20 Years of Emotion Measurement in Advertising". Journal of Advertising Research., 2006, pp.: 18–37.

Sudak, Howard S / Sudak, Donna M: "The Media and Suicide". Washington: Academic Psychiatry. 29(5) 2005, pp.: 495–499.

Young, Charles E.: "Print Ad Research". Ameritest/CY Research, Inc. 2010, Available at: http://www.ameritest.net (12.7.2011)

Andrej Dobrovoljc, Jože Bučar
Faculty of Information Studies
Sevno 13, 8000 Novo mesto, Slovenia
{andrej.dobrovoljc,joze.bucar}@fis.unm.si

Measuring security culture of users of online banking

Abstract: Online banking services offer many benefits compared to the classical ones. They outperform the classic approach in accessibility, comfort, and costs. However, together with comfort, some completely new types of threats have appeared. Intruders have uncovered numerous weaknesses of online services and they are continuously discovering new ones. Online deception is a fact today and we need to find efficient countermeasures. Many studies prove that people are the weakest part in the chain of information security. It is not enough for people to understand the security threats around these modern solutions but they have to behave appropriately as well. Therefore, we are interested in the information security culture of online banking users. In our study, we developed the measuring model for information security culture and applied it on users of online banking in Dolenjska and Bela Krajina. The results prove that the security culture is still not on the sufficiently high level.

Keywords: information security culture, online banking, risk, behaviour, knowledge

1 Introduction

Information technology has drastically improved the quality of our lives on various fields. For example, online banking is one of the most widely used services today and has significantly influenced our daily habits. We can make banking transactions from home or some other location at any time and from various computer devices. Consequently, the vast majority of banks all over the world now offer online banking to their clients. People have more opportunities to choose the most suitable bank for them, since banks are globally present and accessible from almost anywhere. We can conclude that both sides, clients and banks, should profit from it.

The first estimates that the cost of the bank business would drop about 25 % were an exaggeration. In reality, it proved that it is somewhere around 5 %. Today, online banking from a bank's perspective is not an advantage any more but a must. Without it, the bank is unattractive for both potential and existing clients. Therefore, it seems that clients profited more from availability of online banking services.

Online banking for individuals in Slovenia started in the year 2000 and until the end of 2002, there were 98.669 registered users. Up to 2006, this number jumped to 351.111 users, a 256 % increase from 2002. The main reasons for the steep growth were the better availability and higher speeds of internet. During these years, users accepted online banking and found it as one of the key services on the internet. In 2013, the number of users was 690.040, a 96,5 % increase from 2006. These facts prove that online banking is widely accepted today and that it is approaching saturation (Bank of Slovenia 2014).

However, online banking also has some serious drawbacks. The classical banking approach was safe because clients made all transaction at the bank counter and ordered transactions to the bank clerk. With online banking, users should be careful and they have to always check whether they are on the genuine banking website. Even more, intruders are proactive. Under the guise of the bank, they send fake messages to users in order to seize their bank assets. Banks are aware of such threats. Therefore, they introduce various security measures and inform their clients about the importance of secure behaviour. The question is how clients perceive security risks and how they obey the rules of secure behaviour. It is very dangerous if someone understands the security risk and importance of security measures but he or she does not behave accordingly. In this situation, we can speak about the inappropriate information security culture. We define it as the difference between the user's knowledge about security and the actual behaviour of the user.

In the case of online banking, a high information security culture is of utmost importance because people can suffer serious harm. Our intention was to measure the information security culture of users of online banking in Dolenjska and Bela Krajina. We developed a measuring model based on a questionnaire. Our key finding is that the current security culture does not reach a sufficient level.

2 Related work

The information security culture has developed from security culture. It covers the specific environment of informatics, where just the application of various technological security measures is not enough (Rančigaj et al. 2012). The security assurance of the information system (IS) in a great deal depends on the behaviour and participation of people (Mitnick 2002). The key human factors, which support the high information security culture are (Rančigaj et al. 2012):

- perception of importance of security,
- knowledge about security measures, and
- ethics to act according to this knowledge.

The fact is that without the necessary knowledge and the suitable behaviour, we cannot reach the desired security level despite the best possible technological security solutions.

Some researchers (Da Veiga et al. 2010; Niekerk et al. 2006; Ruighaver et al. 2007; Kolkowska 2011) defined their information security culture models above Schein's definition of organisational culture. It consists of three levels:

- **Artefacts** (visible part): behaviour patterns, technology, forms.
- **Values** (partly visible): official documents, which describe values, principals, ethics, and visions.
- **Assumptions** (hidden): people's convictions about how the system works.

Da Veiga and Eloff (Da Veiga et al. 2010) define the information security culture as a set of values, behaviour, assumptions, beliefs, and knowledge of all the IS stakeholders. Their interactions with the IS can be acceptable or inacceptable and they result in a specific form of security assurance, which varies in time. Van Niekerk et al. added the fourth level to the Schein's model, called "information security knowledge" (Okere et al. 2012). This level supports the first three levels and thus ensures compliance.

3 Defining the model

Existing studies focus on measuring the information security culture within the organisation. In the case of online banking, we speak about the culture within the certain group of users. They may not have a direct relationship as it is within the organisation, but they definitely share some experiences of usage in various informal or formal ways. In an organisational way, users are the bank stakeholders and they have to obey the terms and conditions of usage.

A responsible user of online banking has to develop an appropriate attitude toward the usage. In order to achieve a sufficiently high level of culture, he or she has to acquire the knowledge at all levels of information security culture. Therefore, we modified the Niekerk & Von Solms four level model in a way that the knowledge is individually present next to each upper level of security culture. In other words, we doubled the original Schein's model so that we will be able to compare user's behaviour and knowledge on separate levels (Figure 1).

Figure 1: *Information security culture – proposed model extension.*

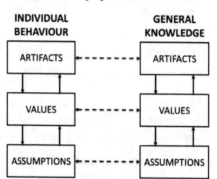

Each behaviour level relates to the corresponding knowledge level. Our intention is to first measure knowledge at the individual level, then to measure related behaviour, and finally to calculate the difference. By our definition, the difference between the level of user behaviour and the level of his or her knowledge represents the security gap. When knowledge is on a desired level and the security gap is small, the information security culture is high. In all the other cases, the security culture is low.

The key difference between knowledge and behaviour is action. People with the same level of knowledge can react differently. Therefore, we have to measure both constructs separately. Questions related to the constructs at the same culture level are similar (Table 1). The only difference is the focus of questions regarding behaviour and knowledge. Questions regarding knowledge reveal the current level of user competences to us, while questions regarding behaviour reveal how seriously users take the knowledge about security.

Table 1. *Focus of questions for measuring the information security culture of users*

Information Security culture levels	User knowledge and perception of security	User behaviour, values, and assumptions
Artefacts	How should we protect?	How do I protect?
Values	What should we protect?	What have I protected?
Assumptions	Why should we protect?	Why do I protect / not protect?

User knowledge about security and its proper perception are the key factors to achieve a high information security culture. However, the service provider has to invest into the awareness and education of clients about general principles of safe usage in order to achieve desired safety.

4 Method

The main purpose of the study was to measure the information security culture of online banking users in Dolenjska and Bela Krajina. We define that the information security culture of online banking users is high when the knowledge about security and user's behaviour are on average above 80 % of the maximum possible value. This definition follows the Pareto principal, which says that we can achieve the result of 80 % with an effort of 20 %. The question is if the banks and the whole society have made enough efforts to achieve a high information security culture of online banking users. In this sense, we place the following hypothesis:

H1:   The information security culture in Dolenjska and Bela Krajina does not reach a sufficiently high level, which is 80 % of the maximum possible value (measuring separately the average of knowledge and the average of behaviour).

H2:   The information security culture is better in Dolenjska than in Bela Krajina.

H3:   The knowledge about security is higher on average compared to people's behaviour.

The measurement instrument is questionnaire based. There is one question for each construct from the proposed model and a five-step Likert scale is used. At the end of the questionnaire, we added some demographic questions. We carried out an online survey. The sample was selected by a snowball method.

5 Results and discussion

In the survey, 260 people participated; however, only 138 questionnaires were completely and correctly filled out. Partially completed questionnaires were removed from the statistical analysis. Among participants, 63 % were male and 37 % were female. Some participants refused to answer the demographic question related to their education level. The education level of participants in the sample is presented in Table 2 and the age structure is presented in Figure 2.

Table 2 Education level of participants

Education	Share
Less than four years of middle school	5 %
Middle school (four years)	42 %
Upper school (two years)	15 %

Education	Share
Bachelor's degree (1. Bologna level)	10 %
Bachelor's degree (before Bologna), Master degree (2. Bologna level)	21 %
Master degree (before Bologna)	3 %
Doctoral degree	4 %

The education structure matches with the statistical data for Slovenia in 2013. Figure 2 proves that online banking is not just a domain of younger people. The majority of users belong to the age group of 41–50 years old. However, older people (51–60 years old) started using online banking as well, despite the fact that IT is something new for them and demands learning. Participants came from different regions. Among them, 65 were from Dolenjska and 51 were from Bela Krajina. Others participants were from other regions.

Figure 2 Age structure of participants in the sample.

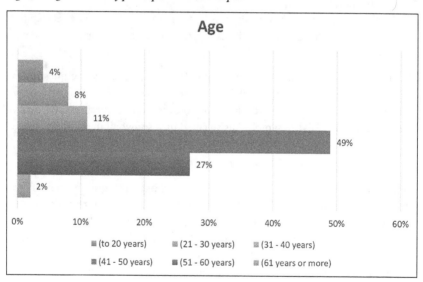

The results of individual concepts from the model are shown in Table 3 and the gap between the user's knowledge and their behaviour is presented in Figure 3. The validity of the hypothesis was tested with Z-statistics (comparing average values) and $\alpha = 0{,}01$.

Table 3 Average values of individual security culture components.

Information Security culture levels	User knowledge and perception of security	User behaviour, values, and assumptions
Artefacts (How?)	3,6	3,5
Values (What?)	3,8	3,7
Assumptions (Why?)	3,7	3,2

Figure 3: Gap between the knowledge and behaviour of online banking users.

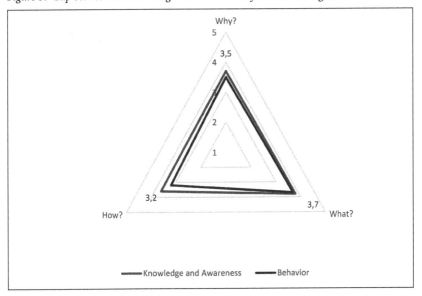

The results prove that the information security culture of online banking users in Dolenjska in Bela Krajina does not reach the suitable level of 80 % (on average at least four on the Likert scale). The knowledge part of the model shows better results; however, none of the components reach the value four. On average, the level of users' knowledge reaches 74 %, while the level of behaviour is at 69 %. We can conclude that users know more about security of online banking, but they behave worse. There is still quite a big gap in expected knowledge (6 %), as well as in behaviour (11 %). Apparently, banks and society have not done enough to achieve sufficient safety within online banking users.

By comparing results in both observed regions, we cannot accept the hypothesis that the information security culture differs among users from Dolenjska and Bela Krajina. This hypothesis was placed due to our assumption that people from rural regions are less educated in IT and its security issues. The comparison proved that this is not the factor, which would influence the information security culture.

6 Conclusion

This study was focused on information security culture of online banking users. As it is one of the most widely used IT services today, and due to its attractiveness for intruders, it is important for information security culture to be high. We analysed the level of security knowledge of users and their behaviour when using online banking. For this purpose, we proposed a new measuring model. The results prove that the information security culture in Dolenjska and Bela Krajina is still not sufficiently high. There is still some lack of knowledge about security, and besides that, users do not behave safely.

In future work, we are going to improve the measurement instrument in a way that we will measure individual concepts of the model with more questions. Besides that, we are going to focus on other interesting online services like Facebook and alike.

7 Acknowledgements

Work supported by Creative Core FISNM-3330-13-500033 'Simulations' project funded by The European Regional Development Fund of the European Union. The operation is carried out within the framework of the Operational Programme for Strengthening Regional Development Potentials for the period 2007–2013, Development Priority 1: Competitiveness and research excellence, Priority Guideline 1.1: Improving the competitive skills and research excellence.

8 References

Bank of Slovenia: Monthly Bulletin. 23 March 2014, p. 187.

Da Veiga, A. / Eloff, J.H.P.: "A framework and assessment instrument for information security culture". *Computers & Security*, 29(2), 2010, pp. 196–207.

Kolkowska, Ella: "Security subcultures in an organization – exploring value conflicts". In *19th European Conference on Information Systems, ECIS 2011. Helsinki. Finland. June 9–11, 2011.*

Mitnick, K.D. / Simon, W.L.: *The Art of Deception: Controlling the Human Element of Security* 1st ed.. New York. USA: John Wiley and Sons, Inc., 2002.

Niekerk, J. Van / Solms, R. von: "UNDERSTANDING INFORMATION SECU-
RITY CULTURE". In *Proceedings of the ISSA 2006 from Insight to Foresight
Conference*. 2006.

Okere, I. / Niekerk, J. Van / Carroll, M.: "Assessing information security culture:
A critical analysis of current approaches". *Information Security for South Africa
(ISSA)*, 2012.

Rančigaj, K. / Lobnikar, B.: "Vedenjski vidiki zagotavljanja informacijske varnosti :
pomen upravljanja informacijske varnostne kulture", 2012, pp. 1–12.

Ruighaver, a. B. / Maynard, S.B. / Chang, S.: "Organisational security culture:
Extending the end-user perspective". *Computers & Security*, 26(1), 2007,
pp. 56–62.

Jana Suklan
School of Advanced Social Studies in Nova Gorica
Gregorčičeva 19, 5000 Nova Gorica, Slovenia
jana.suklan@fuds.si

Vesna Žabkar
Faculty of Economics, University of Ljubljana
Kardeljeva ploščad 17, 1000 Ljubljana, Slovenia
vesna.zabkar@ef.uni-lj.si

Modelling Synergies between Online and Offline Media

Abstract: Advertising spending raises important questions about what is the optimal schedule for multimedia activities. In general, marketers have only limited information about the relative effectiveness of advertising. Thus, information about the effects of online, offline, and their synergetic effect is lacking. Due to high investments in advertising, marketers strive for an analytical tool to measure the advertising effectiveness of different media. Using aggregated market data on advertising and sales, we calibrate an analytic model to establish the presence of synergy between television and internet advertising. An improvement in modelling synergy is enabled by a detailed data collection technique that is tracking sales separately for each media. The model provides insights into how the optimal budget allocations may vary with the presence of synergy between advertising in online and offline media.

Keywords: synergy, media planning, advertising models

1 Introduction

Cross-media synergy, defined as a combined effect of more than one media which exceeds their individual effects on the outcome measure of total sales, is of particular interest (Chang 2004, p. 75; Havlena et al. 2007, p. 215; Naik and Raman 2003, 375). Understanding cross-media synergy is important since is it likely to affect the allocation of resources.

Our focus is on the improvement of existing models that incorporate synergy between media. Our starting point is the extension of the dynamic interactive model using the multiple regression estimation technique. Our observations represent different intervals of time, equally spaced. Through the analysis, based on the detailed collection of reliable proprietary data for each media, we want to contribute to the understanding of consumer response to cross-media strategies.

2 Theoretical background

Managers recognise that consumers combine information they receive from various media, which leads to integrated messages across all media (Raman and Naik 2006, pp. 432–435). The objective of cross-media communication is to increase the final effect, such as sales and brand recognition, by applying various communication channels. Cross-media strategies apply all available communication channels to communicate with the potential customers' in order to maximise cost-efficiency through unified messages creating a consistent image of the company. Effects of individual channels are not independent. Cross-media communications provides for the consolidation of advertising channels, leading to the emergence of synergy (Naik and Raman 2003, pp. 376–384).

Much has been done to try to model synergy in traditional media using common advertising models. Table 1 shows a brief historical evolution from a simple first order autoregressive advertising model introduced by Palda (1965, pp. 164–179) to the more complex ones. Early studies have been conducted in terms of static models based on generalised least square regression (Gatignon and Hanssens 1987, pp. 250–256). Later, models were redefined by adding time components and other modifications using nonlinear square regression or Kalman filter estimation (Naik, Prasad and Sethi 2008, pp. 130–136; Naik and Raman 2003, pp. 379–382). The latest model also includes an interaction term between advertising channels and lagged sales instead of only including the interaction between advertising channels (id, p. 378).

Table 1: Evolution of marketing-mix models

Description	Model
first-order autoregressive advertising model	$S_t = \alpha + \beta_1 u_t + \lambda\, S_{t-1} + v_t$
effects of multimedia advertising	$S_t = \alpha + \beta_1 u_t + \beta_2 v_t + \lambda\, S_{t-1} + v_t$
interaction term	$S_t = \alpha + \beta_1 u_t + \beta_2 v_t + \kappa\, u_t v_t + \lambda\, S_{t-1} + v_t$
carryover by each medium	$S_t = \alpha + \beta_1 u_t + \beta_2 v_t + [\lambda_1 w_1 + \lambda_2(1 - w_1)]\, \lambda_1 S_{t-1} + v_t$

Description	Model
interactions between advertising media and lagged sales when brand sales available by each medium	$S_t = \alpha + \beta_1 u_t + \beta_2 v_t + \lambda S_{t-1} + \kappa_1 u_t v_t + \kappa_2 u_t S_{t-1} + \kappa_3 v_t S_{t-2} + v_t$

Source: Naik and Raman (2003, pp. 376–378)

The purpose of our work is to include the improvement and adjustment of the base model to the observed marketing problem. The improvement in modelling synergy is enabled by separately tracking sales for each media. Doing so, we are allowed to extend the model to lagged sales for each media and explain their effect on total sales. Previous models, having only total sales available, did the estimation of lagged-sales coefficients through maximising the log-likelihood function (ibid.).

3 Empirical analyses

In this section, we describe the data set and present parameter estimation results and assess the effectiveness of the proposed model. While fitting the model to the data, we checked for the existence and implications of the synergy between channels.

3.1 Data Description

The analysis refers to an established brand of high involvement household products. The advertising data set for the brand comprises a daily time series, starting from 2007 to the first half of 2010. The market data consists of television and online sales measured in value of units sold. The proprietary data consists of the number of commercial minutes (for advertising on air) on television in the national market, while online activities are expressed in number of online visitors. In order to maintain confidentiality, we are not in a position to reveal any of the proprietary data, including the identity of the company or its basic products. All the data were previously rescaled and standardised.

Once all fixed-effects are used in the model specification, we constructed a dummy variable (τ_t) to capture the week and weekend effect. The dummy variable captures the fluctuations in demand that is not explained by the advertising data and lag sales. Since we have the empirical information from the marketers about advertising decay and its effect on sales, we also apply it to the 'advertising carry-over' effect, setting it to a two week time span. Furthermore, in practice, it has been observed that if advertising on television terminates, a decline in bookings over the internet is observed after two weeks of interruption.

3.2 Parameters estimation

An effective parsimonious model is what a researcher strives for when building a model. A problem occurs when one already has a large number of independent variables, some of which have multiple components (such as lagged values of advertising or temporal and contextual reference prices). When this occurs, testing out all sorts of interactions can get quite complex. At this point, we need to introduce a term to describe the prolonged or lagged effect of advertising on consumer purchase behaviour. The carryover effect of advertising is defined as a portion of its effect that occurs in time periods following the advertising (Tellis 2007, pp. 507–510).

First, we start by fitting the data to the existing models previously described. For the baseline model, we choose Naik and Raman's model (2003, p. 378) and modify it according to the specifics of data collection. To this end, we introduced our dynamic model, which is an extension of the previous (ibid.) model:

$$S_t = \alpha + \beta_1 u_t + \beta_2 v_t + \kappa u_t v_t + \lambda_1 S_{t-14}^y + \lambda_2 S_{t-14}^u + \tau_t + v_t \qquad \text{(eq. 1)}$$

where κ represents synergy, that is combined sales impact of (u, v) that exceeds the sum of the independent effects $(\beta_1 u + \beta_2 v)$, while lag sales represent a carry-over effect. We denote the sum of sales with S_t, television advertising expressed in minutes with u_t, and online activities expressed in the number of visitors per day with v_t. A dummy variable (τ_t) is used to capture the week and weekend effect.

Table 2: Proposed model

Model parameters	Estimate	Std. Error	t value
Effectiveness of TV advertising, β_1	-0.017	0.055	-0.3
Effectiveness of internet advertising, β_2	-0.087	0.066	-1.3
Synergy between TV and internet, κ	0.24	0.074	3.3
TV carryover effect, λ_1	0.226	0.026	8.6

*the better is the model associated with the smallest value

In eq.1, sales depend on sales of the prior period and all the independent variables that caused prior sales plus the current values of the same independent variables. Our aim is to test the model theoretically and empirically against previously introduced models. We keep all the components (important variables) in the model all the time to avoid a correlation with the error term. A linear model that we are using can also be referred to as a Koyck model, since it includes the lagged dependent variable as an independent variable (Tellis 2007, pp. 514–515). We selected the most parsimonious model (see Table 2). The goodness of fit indicates that the proposed model performs the best according to AIC and BIC. From the proposed model, we can deduce that both the television carryover effect and internet carryover effect play a significant role on sales. This could be explained by the non-instantaneous buying decision related to the high involvement household product. Since a period for reflection is needed, lagged sales (of 14 days) are more exploratory.

Synergy between television and internet is also present and significant at the 95 % confidence interval. According to our data, we can conclude that there is an interaction (synergy) present between advertising on television and the internet. Since television advertising also communicates the brand web page address, it is expected that the TV viewers will visit the web page but not vice versa. That could be a plausible reason why when explaining a model that includes interaction, we have a negative effect on sales, both online and offline. Weekly days also have a statistically significant effect on explaining sales. We are treating weekdays and weekends separately because of the decrease in sales during the weekends.

4 Marketing implications

High positive synergies encourage an increased optimal advertising budget and more investment in otherwise less effective medium. Simple advertising models suggest that in advertising, as the effectiveness of an activity increases, managers should increase the spending on that activity, thus increasing the total media budget. Furthermore, the total budget should be allocated to multiple activities in proportion to their relative effectiveness.

On the other hand, in marketing-mix strategies, it has been proven that if synergy increases, managers should decrease the proportion of the media budget allocated to the more effective communication activity and vice versa. There are several other implications that come with the presence of synergy in the model. The lagged sales term for example is added to track the carryover effect of the sales. So we need to advise managers that if the increase in carryover effect is experienced, the media budget must be increased too. Furthermore, as carryover increases, managers should decrease the proportion of the budget allocated to the more effective activity and vice versa. The counterintuitive part of the proposition is explained through optimal spending which depends not only on its own effectiveness, but also on the spending level for the other activity (Raman and Naik 2006, pp. 435–436).

According to advices followed by previous research on budgeting in the presence of synergy, we can make the same derivations and analyse optimal budgeting between television and internet advertising. Since internet advertising is less effective (we focus on the first simplest model), we could advise managers to decrease the proportion of the media budget allocated to television communication. As we proved the existence of the carryover effect, we can advise that if the increase in carryover effect is experienced, the total budget invested in advertising must be increased too.

We must also emphasise that the optimal budget required for managing advertising activities in the presence of synergy is greater than that required in its absence (Naik and Raman 2003, p. 382).

Anomalies in the models could be explained through the following two segments of consumers: on one side, there are TV viewers who presumably regularly consult information on the internet, on the other hand, there are people who only watch television and do not use nor buy through the internet. The ones that come to light applying synergy are mostly the ones that watch television and consult the internet for further information on a desired product.

5 Discussion

The purpose of this paper is to determine if and how synergies in advertising affect sales through online and offline channels. Using market data, we calibrated the proposed model to establish the presence of synergy between television and internet advertising.

By recognising interaction effects between online and offline activities, managers could be advised to consider inter-activity tradeoffs in optimally planning marketing-mix strategies (Naik, Raman and Winer, 2005, pp. 25–34). Understanding a pattern in consumer reactions to advertising is the path that leads to success and changes the role or marketers. When using a cross-media strategy, planning that takes into account past feedback information and avoids repeating the same mistakes is important. Understanding which variables are important and how variables in the model interact provides the tools to increase the success rate. Marketers' intuitive knowledge must be supported analytically through development of analytical models to capture the essence of a context and limiting the complexity of the model at the same time (Coughlan et al. 2010, pp. 318–320).

List of references

Chang, Yuhmiin / Thorson, Esther.: "Television and web advertising synergies". *Journal of Advertising* 33(2), 2004, pp. 75–84.

Coughlan, Anne T. /Choi, Chan S. / Chu, Wujin / Ingene, Charles A. / Moorthy, Sridhar / Padmanabhan, V. / Raju, Jagmohan, S. / Soberman, David A. / Staelin, Richard / Zhang, John Z.: "Marketing modeling reality and the realities of marketing modelling". *Marketing Letters* 21(3), 2010, pp. 317–333.

Gatignon, Hubert / Hanssens, Dominique M.: "Modeling marketing interactions with application to salesforce effectiveness". *Journal of Marketing Research* 24(3), 1987, pp. 247–257.

Gopalakrishna, Srinath / Chatterjee, Rabikar: "A communications response model for a mature industrial product: Application and implications". *Journal of Marketing Research* 29(2), 1992, pp. 189–200.

Havlena, William / Cardarelli, Robert / de Montigny, Michelle: "Quantifying the isolated and synergistic effects of exposure frequency for TV, print, and Internet advertising". *Journal of Advertising Research* 47(3), 2007, pp. 215–221.

Naik, Prasad A. / Raman, Kalyan / Winer, Russell S: "Planning marketing-mix strategies in the presence of interaction effects". *Marketing Science* 24(1), 2005, pp. 25–34.

Naik, Prasad A. / Raman, Kalyan: "Understanding the impact of synergy in multimedia communications". *Journal of Marketing Research* 40(4), 2003, pp. 375–388.

Palda, Kristian S.: "The Measurement of Cumulative Advertising Effects". *The Journal of Business* 38(2), 1965, pp. 162–179.

Prasad, Ashutosh /Sethi, Suresh P.: "Integrated marketing communications in markets with uncertainty and competition". *Automatica* 45(3), 2009, pp. 601–610.

Raman, Kalyan / Naik, Prasad A.: "Integrated Marketing Communications in Retailing". In: Krafft, Manfred / Mantrala, Murali K. (ed(s).): *Retailing in the 21st Century: Current and future trends*. Springer – Verlag: Berlin Heidelberg 2006, pp. 429–444.

Tellis, Gerard J.: "Modeling Marketing Mix". In: Tellis, Gerard J. / Tim Ambler (ed(s).): *Handbook of marketing research*. Sage Publications: London, UK 2007, pp. 506–522.

Jernej Agrež
Faculty of Information Studies
Ulica talcev 3, 8000 Novo mesto, Slovenia
jernej.agrez@fis.unm.si

Miroslav Bača
Faculty of Organization and Informatics
University of Zagreb
miroslav.baca@foi.hr

Nadja Damij
Faculty of Information Studies
Ulica talcev 3, 8000 Novo mesto, Slovenia
nadja.damij@fis.unm.si

An open framework biometric system optimisation

Abstract: The purpose of this research is to design an open framework optimisation solution for the loosely coupled and uncertain systems applicable to the biometric system. To be able to design an ontology model, we reviewed state-of-the-art biometric systems. Furthermore, we translated the model into relational databases that served as a data input for the optimisation solution. The solution operates as an open framework, tailored to the biometric system. It is capable of responding to its changes and developments and still provides a user with relevant support, either when operating with an existing biometric system or when designing a new multimodal biometric system.

Keywords: loosely coupled system, uncertainty, biometric system, optimisation

1 Introduction

Loosely coupled systems operate with uncertainty. Therefore, optimisation solutions applicable for such systems are an interesting research challenge. Even though the field of organisational optimisation has already been studied in detail, uncertain environments remain a topic that has not been studied to the full extent. In this paper, we present a principled approach to the biometric system, which can easily be included into the family of loosely coupled and uncertain systems. On the first hand, many authors studied biometry and integrated different biometric methods into the system that would cover the specific needs of recognition, control, etc. On the other hand, biometric system's optimisation remains an open research topic.

With the principled optimisation solution, we try to reach several objectives. We try to develop an approach that would create the possibility to model an uncertain system in a transparent, intuitive, and open framework that is capable to capture such system, respond to its changes, and support the user with the optimisation solution. We follow the objectives trough the three-tier optimisation that consists of an open ontology framework, the transformation of the ontology into relational databases, and finally, processing data with the optimisation tool.

2 Loosely coupled and uncertain system optimisation

Loosely coupled systems are widely used conceptually and diversely understood (Orton / Weick, 1990). Considering Glassman's definition of such system, it is clear that it operates with uncertainty, which is reflected through the reduced possibility to foretell detailed descriptions of characteristics of the task that we will attend to solve by the application of the system (Glassman, 1973). Moilanen finds loose coupling and the uncertain operational environment as a stimulus for organisational learning (Moilanen, Sinikka, 2011). As such, the learning process creates a possibility to constantly upgrade system's efficiency, with the support of the optimisation tools. Li et al. argue that the crucial factor when conducting system optimisation is to determine optimal process parameters (Li, 2013). It is possible to achieve such determination through the learning process that can gradually turn from human based learning to automatised decision making. Liu et al. suggest a stepwise approach to system optimisation, starting with the definition of the problem and its characteristics (Liu, / Hong Yi / Zhong-hua Ni, 2013).

We must consider what resources we would use to conduct a problem solving process and consider which resources are the most suitable for the practical application. To be able to proceed with the selection, it is important to link all the resources into a graph and weight important links that represent relations among connected characteristics. Malakooti also finds the ability to evaluate and rank different possible approaches to the problem solving solution to be important, providing us with the insights which are still acceptable among suitable system processes, even though they do not demonstrate the highest suitability (Malakooti, 2012). With such measures, the optimisation process becomes transparent and progressive, which is of high importance when we are dealing with loosely coupled systems, due to their uncertainty and the high possibility of an unexpected change.

2.1 Biometric system

Benziane and Benyettou describe biometrics as a method for the identification of the individual person based on numerous characteristics. They find a biometric system as a highly reliable and fast identification solution (Benziane / Abdelkader, 2011). Dass et al. noticed the application of biometric systems in numerous fields within the social environment, such as: travel, transportation, border control, homeland security, healthcare, banking and finance, access control, airport security, law enforcement, automotive, cyber security, encryption, nuclear power plants, and watermarking (Dass / Karthik / Anil, 2005). Elliot argues that different biometric approaches create the need for a different type of technology to such an extent, that the technology roadmap is needed to be able to understand all the technological aspects and the possibilities within the field of biometry (Elliot, 2005). Following the work of Schatten, we can present the system as the list of methods that consist out of statistical, physical, biological, and other laboratory approaches, on the one side. On the other side, we can find biometric characteristics, such as DNA, body odour, voice, signature, etc. Every included characteristic is also defined by the parameters such as acceptability, feasibility, permanence, measurability, etc. These parameters hold three qualitative values: low, middle, and high. Most of the characteristics are submitted to quality control, recognition, and the structure extraction process. For each of these processes, we use different methods depending on the characteristic itself (Schatten, 2007). Considering that the assessed biometric system consists of 59 methods and 33 behavioral and physical characteristics that differ by 69 types of structures and twelve qualitative parameters, not every biometric engineer has the ability to follow the technology roadmap, especially due to high costs. That is why it is of high importance to have the ability to optimise the system.

2.2. Biometric system optimisation

To be able to approach the biometric system optimisation, we developed a three-phase process. In the first phase, we answered the need to visualise the biometric system and put it into a transparent and easy to understand framework. Based on the theoretical insights we developed an OWL language based ontology, presented in Figure 1.

Figure 1: Ontology of the biometric system

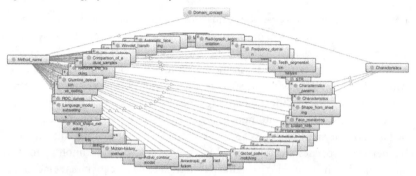

We established the domain concept out of biometric characteristics and biometric methods. As presented in Figure 2, we further divided characteristics into characteristics' parameters, characteristic parameters' values, characteristics' structure, and characteristics' title. The second part of the domain concept is the methods, presented in Figure 3.

Figure 2: Characteristics in the biometric system

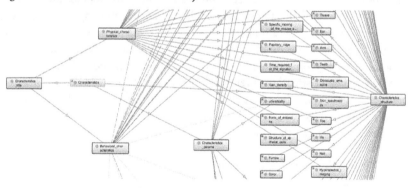

Figure 3: Methods in the biometric system

We selected characteristics, defined their parameters, and connected them with methods, using a literature review as the primary information source. Having included all necessary elements of the biometric system, we needed to determine how to define links between the ontology elements and how to integrate relations based on the links. We decided to use characteristics' title as a backbone of the linking process because the title in most cases relates to all other elements within the ontology structure. Using ontology subclass as a frame for relating the elements, we connected every characteristic title with the type of structure used for biometric purposes, with a feasible method for recognition, structure extraction, and for quality control. We linked the characteristic with the behavioural or physical attribute and further linked it with the known assessment parameters according to the theoretical insights from the literature review.

To exceed bare linking, we added the weight that served as a relation indicator to selected links. When connecting characteristics with the available methods, in many cases, there is more than one feasible method. To reveal such cases, we linked characteristics to the method through the relation "some". If the characteristic used a single method for a specific purpose, we established a link with the relation "only". Another relation that we needed to integrate into the link frame was the "high", "middle," and "low" values that defined a set of characteristic parameters. We set the properties again as the subclass of the characteristic, using the same type of "some" and "only" relation to reveal which values correspond to which characteristic parameters.

In the second phase of the optimisation process, we transformed ontology from its OWL form into five relation databases, where we included previously defined relations. The most comprehensive database presents relations between characteristics and their parameters together with their structure. The second database consists of the characteristic parameters, defining where high, medium, and low values are labeled as positive, neutral, or negative. The last three databases contain relations between characteristics and methods for quality control, recognition, and structure extraction.

The third and most important phase in the optimisation process was the design of a tool that would support user decision making, when facing the need to select the most appropriate methods out of the biometric system. According to the importance of designing the tool as an open framework, we choose to design it in an R programming environment. The tool proceeds with the optimisation in four steps. The importance of such procedure is the possibility of the user to quickly and simply add additional characteristics, methods, and other elements to the data, if needed. In the first step, the user selects the data source and relation databases. In the following step, the user defines the ten most appropriate biometric characteristics that he or she operates with or that will take an important role in his or her biometric based solution. The third and the most comprehensive step consists of pondering the ten chosen characteristics on a scale from 0.00 to 1.00. The pondering process is based on the following criteria.

When all inputs are selected, the tool first calculates how many positive, neutral, and negative parameters are held by a single characteristic. Further on, it calculates the number of characteristic's structures and counts the number of methods used for quality control, recognition, and structure extraction. All the calculated values are then recorded in the data frame created by the tool. When the data frame is created (Table 1), weights are calculated for a single instance, adding 0.1 points to the weight if the characteristic has more positive parameters than neutral, if the characteristic has more positive parameters than negative, and when it has more neutral parameters than negative. Further, 0.1 points are added to the characteristic's weight when its number of structures, structure extraction methods, quality control methods, and recognition methods exceed the mean of the structure number within the data frame.

Table 1: Example of the data frame, used for optimisation

positive	neutral	negative	structures	quality	recognition	extraction	weight	no
0	0	2	1	0	0	0	0.00	1
2	5	4	1	6	2	4	0.40	2
4	4	4	4	0	0	1	0.10	3
5	4	3	5	1	0	2	0.50	4
4	1	2	2	1	0	2	0.40	5
3	0	0	0	0	0	0	0.20	6
1	5	5	1	0	0	1	0.00	7
4	5	2	2	4	1	5	0.60	8
1	4	2	0	0	0	0	0.10	9
4	1	7	1	7	1	3	0.40	11

In the fourth step, the tool arranges weights on the scale from the lowest to the highest, revealing the value of the weight and serial number of corresponding characteristics to the user.

3 Conclusion

This paper includes insights about the optimisation of the biometric system, which, due to its uncertainty, holds characteristics of a loosely coupled system. The uncertainty of the biometric system follows the state, where we implement systems that are hard to predict. At the same time, it includes several different criteria that define how to apply it in order to achieve the most optimal results. The main purpose of this research was to develop an optimisation solution that would operate as an open framework and that could ensure the selection of the most appropriate biometric approaches according to the given situation. To be able to set up such a solution, we decided to design it in three levels. At the first level, we created ontology, where we integrated all the elements of the biometric system and linked them according to the known rules and characteristics of the biometric system. Ontology provided us with a transparent overview of the biometric system, allowing us to observe it from a holistic perspective, as well as to dive into the specific relations among different elements of the system. On the ground of the ontology, we translated the systems' elements and their relations in the several relation databases and developed optimisation tool within the R programming environment. An optimisation tool operates on the ground of the user's selection of the biometric characteristic, and further on allocates them on the scale from 0.00 to 1.00. Allocation is assigned from the weights, based on the evaluated parameters of biometric characteristics, quality control, recognition, and structure extraction methods that are feasible with the selected characteristics.

We divide major conclusions of our work into two general aspects. A first aspect highlights ontology as the primary stage of the systems' optimisation. Ontology provides us with knowledge about the system and its elements. According to high uncertainty of the biometric system, it is necessary to model its structure within the ontology framework that is capable of linking its elements through the several different characteristics and other predefined connections, while being able to include any necessary change. A second aspect exposes the importance of the smooth downgrading conversion from the knowledge retained in the ontology into the data retained in the relation databases. It is necessary to proceed with such operation due to the need to apply optimisation based on measurements of system characteristics.

When we operate with the data that describes the system in detail, the optimisation tool becomes a framework that should be designed in an open manner as well. Openness is an attribute of high importance that provides us with the ability to include all necessary changes, originating on the first hand, from the change in the system, and on the other hand, from the additional optimisation needs.

For the future research, we recommend design of the solution that would be capable to conduct direct transformation from the OWL (XML) form ontology to the CSV form relation databases. At the same time, the existing optimisation tool does not provide the user with the ability to define his or her own pondering criteria, which would importantly increase usability of the optimisation tool.

4 References

Benziane, Sarah / Abdelkader Benyettou: "An introduction to Biometrics". *International Journal of Computer Science and Information Security* 9(4), 2011

Dass, Sarat C. / Karthik Nandakumar / Anil K. Jain. "A Principled Approach to Score Level Fusion in Multimodal Biometric Systems". *Audio-and Video-Based Biometric Person Authentication*, Springer: Berlin Heidelberg 2005, pp. 1049–1058

Elliot, John: "Biometrics roadmap for police applications". *BT Technology Journal* 23(4), 2005, pp. 37–44

Glassman, Robert B: "Persistence and Loose Coupling in Living Systems". *Behavioral Science* 18(2), 1973, pp. 83–99

Li, Shujuan / Yong Liu / Yan Li / Robert G. Landers / Lie Tang: "Process planning optimization for parallel drilling of blind holes using a two phase genetic algorithm". *Journal of Intelligent Manufacturing* 24(1), 2013, pp. 791–804

Liu, Xiao-jun / Hong Yi / Zhong-hua Ni: "Application of ant colony optimization algorithm in process planning optimization". *Journal of Intelligent Manufacturing* 24(1), 2013, pp. 1–13

Malakooti, Behnam: "Decision making process: typology, intelligence, and optimization". *Journal of Intelligent Manufacturing* 23(1), 2012, pp. 733–746

Moilanen, Sinikka: "Learning and the loosely coupled elements of control". *Journal of Accounting & Organizational Change* 8(2), 2011, pp. 136–159

Orton, J. Douglas / Weick, E. Karl: "Loosely coupled systems: A reconceptualization". *The Academy of Managerial Review* 15(2), 1990, pp. 203–223

Schatten, Markus. Zasnivanje otvorene ontologije odabranih segmenata biometrijske znanosti. (master thesis) Fakultet organizacije i informatike: Varaždin 2007

Biljana Mileva Boshkoska
Faculty of Information Studies
Sevno 13, 8000 Novo mesto, Slovenia
Biljana.mileva@fis.unm.si

A framework for qualitative evaluation of air pollution levels

Abstract: Air pollution is a constant environmental problem, which introduces significant costs primarily to the health of society, ecosystems, and the economy. The main difficulty in modern modelling approaches is their interpretation by the wider public as well as dissemination of the information of air pollution levels, mainly because of the complex modelling procedures. In this paper, we provide a preliminary modelling approach that can be used for the aggregation of the different air quality pollutants into one air quality measure described qualitatively and hence can be understood by the wider population. The paper proposes the modelling approach DEX that allows for both scientific notation of the model and easy-to-interoperate evaluation results by the wider public.

Keywords: air pollution evaluation, air quality, decision expert modelling

1 Introduction

The media reports may lead us to believe that the problem of ambient air pollution arose in the second half of the last century. However, the pollution of ambient air is not a new phenomenon in the history of humankind. Obviously, even for the caveman, the lighting of fire had its consequences (Elvingson, Agren 2004). There is historical data on destruction of plants because of furnaces since the Roman Empire. However, the type of pollution of ambient air to which people have been exposed to has changed throughout history, but the problem has been known for a long time, and caught the interest of the public, especially in the 14th century when people first began to use coal for heating in their homes (Schnelle, Brown 1997).

Today, the problems due to air pollution exist in several areas. Air pollution is a constant environmental problem, which introduces significant costs to the health of society, ecosystems (Effects of air pollution on European ecosystems, Past and future exposure of European freshwater and terrestrial habitats to acidifying and eutrophying air pollutants, EEA Technical report, No 11/2014 2014), and the economy.

The most important and obvious problems to air pollution are the direct effects on human health. The three pollutants that are recognised to most significantly affect human health are Particulate matter (PM), nitrogen dioxide (NO_2) and ground-level ozone (O_3) (European Environment Agency, Air pollution 2014; Air Climate & Climate Secretariat, Toxic air in Europe 2014). Recent research indicates that small particles that can be caught with filters of 2.5 μm ($PM_{2.5}$) in the air caused about 450,000 premature deaths within the 27 European Union (EU) countries in the year 2005. Another 20,000 premature deaths were caused by ground-level ozone (Air Climate & Climate Secretariat, Air Quality 2014). Small particles were also responsible for around 100,000 serious hospital admissions in the EU25, and a much larger number of less serious effects, such as some 30 million respiratory medication use days and several hundred million restricted activity days.

Next in line are the effects and damage to our environment such as the acidification of lakes, including the soil, deforestation, eutrophication, ozone at ground level, or crop damage. Finally, the problems of air pollution overlap with other complex environmental issues such as congestion and mobility, land use, and global warming.

Today, science is concerned with modelling of parameters of the ambient air in order to investigate or to protect and improve the environment in which we live (Cheremisinoff, 2006), (Boubel, et al. 1994), (Turner, 2004). Many research studies deal with the modelling of air quality in order to fill in the gaps of missing data, or to predict the air quality (Rakotomamonjy, 2003), (Berkowicz, Winther, Ketzel, 2006), (Calori et al., 2006), (El-Halwagi, 1997), (Farias, ApSimon, 2006), (Cabrera et al., 2006), (Raimondo et al., 2007), (Canu, Rakotomamonjy, 2001), (Neofytou et al., 2006), (Sokhi et al., 2006). These models are usually based on historical data; they use complex algorithms and are usually understood by a small population of the relevant domain researchers. The paper aims at providing a tool for determining the air pollution level as an aggregated value. Moreover, the tool should be easily used by a wide population without the need for an expert level understanding of the air pollution processes or modelling techniques. In particular, this modelling approach can be used for real-time data on air pollution levels. The background motivation of this work arises from the fact that web sites usually provide information regarding different pollutant levels, without aggregated information about the overall air pollution. For example, if we check the EEA web site (European Environmental Agency 2014), we get maps for four unvalidated real-time air quality pollutants across Europe:

O_3, NO_2, PM_{10}, and Sulphur dioxide (SO_2). However, there is no aggregated map of the overall air quality.

The paper provides a preliminary modelling framework that can be used for the aggregation of the different air quality pollutants into one air quality measure described qualitatively and hence can be understood by the wider population.

2 Methodology

In order to protect human health and vegetation, the Council directive 1999/30/EC has been adopted which defines the limit values of air pollutants and the dates until which these had to be met. Despite these measures, the air quality in European countries frequently breaches the limits of allowed concentrations of ambient air parameters (From The Atlantic Citylab, Western Europe's Mild March, 2014), (UK Air pollution, 2014). Reasons for increased air pollution can be found in the natural weather conditions as well as in the economic crises (The Christian Science Monitor, How Greece's economic crisis filled the Athens sky with smog 2014).

The limit values of the parameters of ambient air are given in Table 1 – Table 4.

Table 1: Limit value and the alert threshold for sulphur dioxide (Directive 2008/50/EC of the European parliament and of the council of 21 May 2008 on ambient air quality and cleaner air for Europe 2008)

	Averaging period	Limit value micrograms/m³	Comment
Hourly limit value for the protection of human health	1 hour	350	Not to be exceeded more than 24 times a calendar year
Daily limit value for the protection of human health	24 hours	125	Not to be exceeded more than three times a calendar year
Limit value for the protection of ecosystems	Calendar year and winter (1.9–31.3)	20	

Table 2: *Limit value and the alert threshold for* NO_2 *and oxides of nitrogen (Directive 2008/50/EC of the European parliament and of the council of 21 May 2008 on ambient air quality and cleaner air for Europe 2008)*

	Averaging period	Limit value micrograms/m³	Comment
Hourly limit value for the protection of human health	1 hour	200 NO_x	Not to be exceeded more than 24 times a calendar year
Daily limit value for the protection of human health	24 hours	40 NO_2	Not to be exceeded more than three times a calendar year
Limit value for the protection of vegetation	Calendar year	30 NO_x	

Table 3: *Limit value and the alert threshold for particulate matter (Directive 2008/50/EC of the European parliament and of the council of 21 May 2008 on ambient air quality and cleaner air for Europe 2008)*

	Averaging period	Limit value micrograms/m³	Comment
24-hour limit value for the protection of human health	24 hours	50	Not to be exceeded more than seven times a calendar year
Anual limit value for the protection of human health	Calendar year	20	

Table 4: *Target value for ozone (Directive 2008/50/EC of the European parliament and of the council of 21 May 2008 on ambient air quality and cleaner air for Europe 2008)*

	Averaging period	Target value micrograms/m³	Comment
Protection of human health	Maximum daily eight-hour mean	120	Not to be exceeded more than 25 days per calendar year averaged over three years
Protection of vegetation	May to July	AOT40 (calculated from one h values)	

The values of the air pollutants are continuous measurements, which can be converted into qualitative ones by using the guidelines in (23). These conversion levels are given in Table 5:

Table 5: *Quantitative to qualitative mapping of pollutants levels according to guidelines in (Directive 2008/50/EC of the European parliament and of the council of 21 May 2008 on ambient air quality and cleaner air for Europe 2008)*

Level of O3 in micrograms/ m^3	Level of NO$_x$ (NO and NO$_2$) in micrograms/m^3	Level of PM (PM$_{10}$ and PM$_{2.5}$) in micrograms/m^3	Level of SO$_2$ in micrograms/m^3	Qualitative description
0 – 60	0 – 50	0 – 20	0 – 50	Low
60 – 120	51 – 100	21 – 35	51 – 100	Slight
120 – 180	101 – 200	36 – 50	100 – 300	Moderate
180 – 240	201 – 400	51 – 65	301 – 500	High
Above 240	Above 400	Above 65	Above 500	Very high

2.1 Qualitative multi-criteria decision analysis and DEX methodology

The process of decision-making (DM) is an important field not just for computer science, but for science in general. Indeed, DM algorithms are applicable in a wide range of real life situations, including very complex but nevertheless realistic problems.

Multi-criteria decision analysis (MCDA) is a sub-discipline of operations research, concerned with structuring and solving decision problems that involve multiple criteria. For a given set of decision options, MCDA considers three types of problematics: choosing, sorting, and ranking (Roy 2005). Choosing means a selection of one option (or a sub-set of options) from the set of decision options as the best ones. Sorting aims at assigning a class to each of the available options from a set of predefined classes. Ranking aims at defining a complete or partial order on a given set of options.

The problem addressed here is directly motivated by decision expert (DEX) methodology (Bohanec 2013; Bohanec, Rajkovic 1990). DEX is a qualitative modelling methodology that, in the process of developing a decision model, produces decision tables, which are interpreted as either a set of options or a set of decision rules governing the preference evaluation. The specifics of the method is that it does not use numbers, but rather it uses qualitative words, thus allowing structuring of an experts knowledge in a user-friendly and understandable way. DEX has been widely used and explored for decision problems that are understandable by humans and which are described with qualitative attributes.

There are two types of attributes in DEX: basic attributes and aggregated ones. The former are the directly measurable attributes, also called input attributes. Input attributes are directly measured and are given with a qualitative description. Aggregate attributes are obtained by using a utility function for aggregation of the basic and/or other aggregated attributes.

Attributes in DEX form a hierarchical structure that represents a tree. The evaluation of options using the hierarchical tree is obtained by applying a bottom-up approach. This approach implies that values of basic attributes are provided to the leaves of the tree. These values are aggregated into higher levels up to the root of the tree. Finally, the value of the root of the tree represents the evaluation of the option. In the hierarchical tree, attributes are structured so that there is only one path from each aggregate attribute to the higher-level one. The path contains the dependencies among attributes such that the higher-level attributes depend on their immediate descendants in the tree. This dependency is defined by a utility function. The utility functions are expert defined functions. Such utility functions, also called decision tables, are given in Table 6 and Table 7. In the decision table, rows represent options, while columns represent attributes. The last attribute is the aggregated one. Its value is provide by the decision-maker and it expresses his/her preferences (expertise) over the options. Those options that are equally preferred (or evaluated) belong to the same qualitative class.

In the DEX, the aggregation of the qualitative attributes into a qualitative class in each row in the decision table can also be interpreted as "*if-then*" rules. For example, the first row in Table 7 can be interpreted as *if PM_{10} is Low AND $PM_{2.5}$ is Low then PM is Low*. This enables both practitioners and the wider population to make use of the results. Practitioners obtain a scientific representation of the data, while the wider public obtains an easy to understand explanation about the air pollution.

3 Results

In order to build a DEXi model for the presented problem, one should first define the qualitative value scales of each attribute. For the purposes of this paper, the quantitative values of each attribute are divided into five ordered intervals, as presented in Table 5. For example, the values for ozone are divided into intervals of 0–60, 60–120, 120–180, 180–240, and above 240 µgr/m³. This procedure is performed for all attributes. Next, a qualitative description is provided to each of these intervals, as shown in the last column in Table 5. For example, the values of ozone that belong to the interval 0–60 μgr/m³ are described as *low*, the values in the next interval are described as *slight*, etc. Next, a utility table is defined for each of the aggregated attributes resulting from the final evaluation model. The attributes value tree and their qualitative scale values for the evaluation of the air quality levels are presented in Figure 2. The values of the presented scales for each attribute are ordered so that the value presented with the green colour is the most preferred one, while the value in red colour is the least preferred one. In other words, values described with *low* are preferred in this example.

The developed DEX model tree for evaluation of air quality levels is presented in Figure 1. The basic attributes in Figure 1 are given with rectangles with curved edges. Such attributes are $PM_{2.5}$, PM_{10}, NO_2, NO, and O_3. The aggregated attributes are given with rectangles with sharp edges, such as PM, NO_x, O_3, and Air Quality.

Figure 1 DEXi model tree for evaluation of air pollution

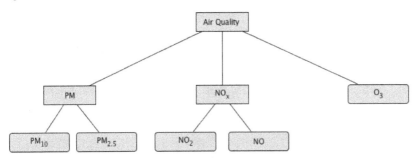

Figure 2 Attribute value tree and scales

Attribute	Scale
Air Quality	*low*; slight; moderate; high; very high
├PM	*low*; slight; moderate; high; very high
│ ├PM10	*low*; slight; moderate; high; very high
│ └PM2.5	*low*; slight; moderate; high; very high
├NOx	*low*; slight; moderate; high; very high
│ ├NO2	*low*; slight; moderate; high; very high
│ └NO	*low*; slight; moderate; high; very high
└O3	*low*; slight; moderate; high; very high

Table 6: Aggregation of the basic attributes NO₂ and NO into the attribute NOₓ

NO2	NO	NOx
50%	50%	
1 *low*	<=slight	*low*
2 <=slight	*low*	*low*
3 *low*	moderate	slight
4 slight	slight	slight
5 moderate	*low*	slight
6 slight:moderate	moderate	moderate
7 moderate	slight:moderate	moderate
8 <=high	high	high
9 high	<=high	high
10 *	very high	very high
11 very high	*	very high

The evaluation of the air quality is performed by using aggregation functions given in a tabular format. For example, the aggregation of the qualitative attributes PM_{10} and $PM_{2.5}$ into a qualitative attribute PM is given in Table 7. Such tables are created for all aggregated attributes presented in Figure 1. These aggregation functions can also be read as *if-then* rules, thus making them understandable for the general public.

Table 7: Aggregation of basic attributes PM10 and PM2.5 in qualitative attribute PM

PM10	PM2.5	PM
50%	50%	
1 *low*	<=slight	*low*
2 <=slight	*low*	*low*
3 *low*	moderate	slight
4 slight	slight	slight
5 moderate	*low*	slight
6 slight:moderate	moderate	moderate
7 moderate	slight:moderate	moderate
8 <=high	high	high
9 high	<=high	high
10 *	very high	very high
11 very high	*	very high

Using the proposed method, an evaluation of the air pollution level is performed for Ljubljana, for aone week in the period of 26.2.2015–5.3.2015. The quantitative values of the measured parameters are obtained from the official web site of the Slovenian Agency for Environment *(Slovenian Agency for Environment 2014)* and are presented in Figure 3. Based on these results, the evaluation of the air pollution level is performed for one weekday, from 26.2.2015–4.3.2015 based on the DEXi model. The results from the evaluation are given in Figure 4.

Figure 3 Daily pollution values for Ljubljana, Slovenia from 26.2.2015–5.3.2015

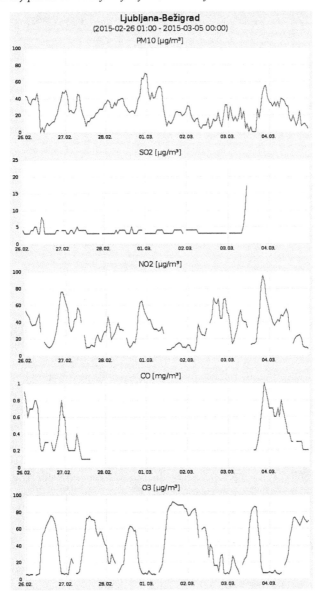

Figure 4: Evaluation of daily air pollution for one week in Ljubljana

Option	26.2.2015	27.2.2015	28.2.2015	1.3.2015	2.3.2015	3.3.2015	4.3.2015
. Air Quality	moderate;high	moderate;high	moderate;high	moderate;l	moderate;high	moderate;high	moderate;high
.. PM	high;very high	high;very high	high;very high	high;very h	high;very high	high;very high	high;very high
... PM10	high	high	high	high	high	high	high
... PM2.5	*	*	*	*	*	*	*
.. NOx	low	low	low	low	low	low	low
... NO2	low	low	low	low	low	low	low
... NO	low	low	low	low	low	low	low
.. O3	slight	slight	slight	slight	slight	slight	slight

Attributes: 8 (5 basic, 0 linked, 3 aggregate) | Scales: 8 | Functions: 3 | Options: 7

4 Conclusions

Air pollution introduces significant costs to the health of people, the ecosystem, and the economy. Therefore, there is a continuous need of improving the air pollution models of ambient air, as well as raising awareness regarding the consequences of the high levels of air pollution. This paper offers a modelling approach of informing the public about the levels of air pollution in both a precise and understandable way.

In particular, the paper provides a framework for the qualitative modelling of the air pollution levels. It is based on the DEX methodology that offers the aggregation of the different air quality pollutants into one air quality measure described qualitatively. The DEX methodology offers both scientific notations of the model and easy-to-interoperate evaluation results by the wider public. Future work should include more environmental parameters in the DEX model, including different air pollutants.

5 Acknowledgements

This work is supported by Creative Core FISNM-3330-13-500033 'Simulations' project funded by The European Regional Development Fund of the European Union.

6 References

Air Climate & Climate Secretariat, Air Quality, retrieved 15.9.2014, from http://www.airclim.org/air-quality,.

Air Climate & Climate Secretariat, Toxic air in Europe, retrieved 15.9.2014, from http://www.airclim.org/toxic-air-europe.

Berkowicz, Ruwim / Morten Winther / Matthias Ketzel: "Traffic pollution modelling and emission data." *Environmental Modelling & Software* 21, 2006, pp. 454–460.

Bohanec, Marko / Vladislav Rajkovic: "DEX: An expert system shell for decision support." *Sistemica* 1, 1990, pp. 145–157.

Bohanec, Marko: *DEXi: Program for Multi-Attribute Decision Making: User's manual: version 4.00*. Ljubljana: IJS Report DP-11340, Jožef Stefan Institute. 2013.

Boubel, Richard / Donald L. Fox / Bruce D. Turner / Arthur C. Stern: *Fundamentals of Air Pollution*. Academic Press. 1994.

Cabrera, González / Fidalgo, Martínez / Martín, Mateos / Vicente, Tavera: "Study of the evolution of air pollution in Salamanca (Spain) along a five-year period (1994–1998) using HJ-Biplot simultaneous representation analysis." *Environmental Modelling & Software* 21 (1), 2006, pp 61–68.

Calori, Giuseppe / Monica, Clemente / Robera, De Maria / Sandro, Finardi/ Francesco, Lollobrigida / Gianni Tinarelli: "Air quality integrated modelling in Turin urban area." *Environmental Modeling & Software* 21 (4), 2006, pp. 467–476.

Canu, Stephane/ Alain, Rakotomamonjy: "Ozone peak and pollution forecasting using support vectors." *IFAC Workshop on Environmental Modelling*. Yokohama, Japan. 2001.

Cheremisinoff, Nicholas P.: *Handbook of Air Pollution Prevention and Control*. 2006Butterworth Heinemann.2006.

Directive 2008/50/EC of the European parliament and of the council of 21 May 2008 on ambient air quality and cleaner air for Europe, retrieved 15.9.2014, from http://eur-lex.europa.eu/

Effects of air pollution on European ecosystems, Past and future exposure of European freshwater and terrestrial habitats to acidifying and eutrophying air pollutants, EEA Technical report, *No 11/2014*, retrieved 15.9.2014, from http://www.eea.europa.eu/publications/effects-of-air-pollution-on.

El-Halwagi, Mahmoud M.: *Pollution Prevention through Process Integration*. Academic Press. 1997.

Elvingson, Per / Christer, Agren: *Air and the environment*. Goteborg, Sweden: Elanders Infologistics AB. Molnlycke. 2004

European Environmental Agency, retrieved 15.9.2014, from http://www.eea. europa.eu/themes/air/air-quality/map/real-time-map.

European Environment Agency, Air pollution, retrieved 15.9.2014, from http:// www.eea.europa.eu/themes/air/intro.

Farias, F. /Helen ApSimon: "Relative contributions from traffic and aircraft NOx emissions to exposure in West London." *Environmental Modelling & Software* 21 (4), 2006, pp. 477–485.

From The Atlantic Citylab, Western Europe's Mild March, retrieved 15.9.2014, from http://www.citylab.com/commute/2014/03/western-europes-mild-march-has-led-air-quality-crisis-france/8640/.

Neofytou, Panagiotis / Alexandros G. Venetsanos / Stylianos, Rafailidis / John G. Bartzis: "Numerical investigation of the pollution dispersion in an urban street canyon." *Environmental Modelling & Software*, 21 (4), 2006, pp. 525–531.

Raimondo, Giovanni / Alfonso, Montuori / Walter, Moniaci / Eros, Pasero / Esben, Almkvist: "A machine learning tool to forecast PM10 level." *Fifth Conference on Artificial Intelligence Applications to Environmental Science*. San Antonio, Texas, USA. 2007.

Rakotomamonjy, Alain: "Variable Selection using Support Vector Machines based Criteria." *Neurocomputing* 55, 2003, 1–2.

Roy, Bernard: "Paradigms and challenges." In *Multi Criteria Decision Analysis: State of the art survays*, 2005, pp. 3–24.

Schnelle, Karl B., and Charles A. Brown. n.d. *Air Pollution Control Technology Handbook*. 1997: CRC Press.

Slovenian Agency for Environment, retrieved 5.3.2015, from http://www.arso.gov. si/en/air/data/amp/e00_g_30.html.

Sokhi, Ranjeet S. / R. San, Jose / Nutthida, Kitwiroon / Evangelia, Fragkou / Juan.L., Perez / Douglas R., Middleton: "Prediction of ozone levels in London using the MM5–CMAQ modelling system." *Environmental Modelling & Software* 21 (4), 2006, pp. 566–576.

The Christian Science Monitor, How Greece's economic crisis filled the Athens sky with smog, retrieved 15.9.2014, from http://www.csmonitor.com/World/ Europe/2014/0212/How-Greece-s-economic-crisis-filled-the-Athens-sky-with-smog.

Turner, Bruce D.: *Workbook of atmospheric dispersion estimates*. Lewis Publishers. 2004.

UK Air pollution, retrieved 15.9.2014, from http://blogs.egu.eu/hazeblog/2014/ 03/14/uk-air-pollution-march-2014/.

Albert Zorko, doc. dr. Zoran Levnajić
Faculty of Information Studies
University of Novo mesto
Sevno 13, 8000 Novo mesto, Slovenia
albert.zorko@gmail.com
zoran.levnajic@fis.unm.si

The overview of recent findings in diagnostics of mental disorders

Abstract: Psychological states have a huge impact on the autonomic nervous system. The literature on this scientific area is far from simple; however, mental states have an effect on many states of the autonomic nervous system. In the following paper, we highlight an overview of important research in this area. We were considering different conceptual approaches which help us to better understand this interesting and very important scientific area for human health from a different perspective.

Keywords: mental disorders, heart rate variable, diagnostics, cardiorespiratory coupling

1 Introduction

Background: A large amount of clinical psychophysiology research deals with autonomic dysfunctions. Current methods only investigate a subordinate autonomic structure instead of the entire autonomic nervous system. In order to achieve a new breakthrough, it is necessary to introduce new methods.

Researchers use deferent methods to evaluate autonomic dysfunction. A very commonly used method is measuring heart rate and respiratory rate variability. The frequency specific fluctuation of heart rate is assessed with power spectrum analysis. Other commonly used methods are: electrocardiography, electroencephalography, blood pressure variability measure, microneurography, mean arterial pressure, and transcutaneous oxygen.

1.1 Autonomic nervous system disorders

The autonomic nervous system is part of the human nervous system. It consists of all nervous pathways leaving the central nervous system that have a ganglionic synapse situated outside the central nervous system. There are three distinct anatomical divisions: sympathetic, parasympathetic, and enteric nervous system (Mai & Paxinos, 2012). The functioning of the autonomic nervous system strongly depends on the equilibrium between the parasympathetic and the sympathetic nervous system. The complex structure of the autonomic nervous system can be seen in Figure 1.

Parasympathetic activation results in a slowing of the heart rate, a reduction in the force of contraction in the atria, and a reduction in conduction velocity through the atrioventricular node (Brading, 1999).

Sympathetic activation results in an increase in the force and rate of cardiac contraction, increased blood pressure from both increased peripheral resistance and increased cardiac output, increased salivation, dilation of the pupils in the eye, increased sweating, piloerection, inhibition of sodium excretion, and trembling (Brading, 1999).

Figure 1: Autonomic nervous system (Blessingm & Gibbins, 2008)

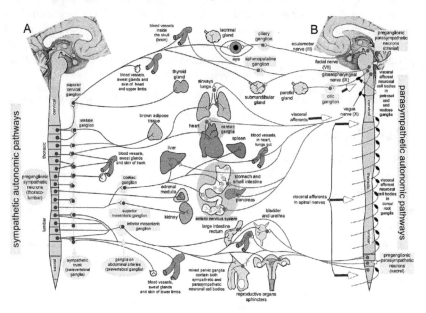

Many factors, including mental state, influence the autonomic nervous system. As an example, we should look at Anxiety, Obsessive Compulsive Disorder, and Post Traumatic Stress Disorder. The real cause is still unknown; however, there are many known triggers for these mental disorders. In addition to external factors, chemical imbalances in the nervous system are very common.

A typical example is excess adrenaline production. Glucose is the forerunner of Biological Energy called Adenosine triphosphate (ATP), which is essential in the manufacture of the relaxing and feel good neurotransmitters, such as serotonin.

When hypothalamic–pituitary–adrenal axis (HPA axis) senses a low blood sugar level, it will send a hormonal message to the adrenal glands to pour adrenaline into the system. This raises blood sugar levels and will feed the brain again, but it also causes us to feel fearful without an external object of fear. There are many reasons for this because there are many medical conditions that interfere with the proper absorption of glucose (Plesman, 2011).

Another example is bipolar disorder (manic-depressive illness). The triggers can be all sorts of life events with positive or negative nature. Illness is associated with changes in various neurotransmitter levels and activity, commonly referred to as a chemical imbalance in the brain (Albrecht & Herrick, 2007).

2 Diagnostic methods

2.1 Classical diagnosis

Classical diagnosis began with the physician asking questions about the medical history, and sometimes with a physical exam. There are no laboratory tests to specifically diagnose mental illness. Doctors use different tests to exclude other diseases. Specially designed interview and assessment tools are used to evaluate a person for mental illness (Goldberg, 2014). The World Health Organization has approved two questionnaires: General Health Questionnaire (GHQ) – 12 and GHQ – 28. Results were uniformly good when using both questionnaires (Goldberg & et al., 1997). Diagnosis becomes very difficult when differential[1] diagnosis must be made.

2.2 Magnetic resonance imaging

Researchers are seeking for alternate diagnostic methods and for a more objective diagnosis. Magnetic resonance imaging (MRI) is a promising method, but not for all mental disorders. Better accuracy is achieved for Attention-Deficit/ Hyperactivity Disorder (ADHD), schizophrenia, Tourette's, and bipolar disorder. More research still needs to be done to ensure success (Marlow, 2012).

Smaller hippocampal volume predicts pathologic vulnerability to severe stress (Gilbertson, 2002). Using MRI, Douglass Bremner and his associates discovered that patients with depression had a statistically significant (19 %) smaller left hippocampal volume than comparison subjects, without smaller volumes of comparison regions (amygdala, caudate, frontal lobe, and temporal lobe) or whole brain volume. The findings were significant after brain size, alcohol exposure, age, and

---

1   Diagnosis for multiple illness

education were controlled for (Bremner & et al., 2000). Some other caveats are: the system will probably struggle with patients who have more than one diagnosis, the system is unable to detect early stages of disorders, and the diagnostic categories might not be biologically valid. (Makin, 2013)

2.3 Electroencephalograph (EEG)

A statistical machine learning methodology on electroencephalogram (EEG) data is used for the diagnosis of psychiatric illnesses: major depressive disorder, chronic schizophrenia, and bipolar depression. The average correct diagnosis rate attained using the proposed method is over 85 %, as determined by various cross-validation experiments (Khodayari-Rostamabad et al., 2010). A novel research ideology, multi-paradigm methodology, and advanced computational models for automated EEG-based diagnosis of neurological and psychiatric disorders was presented at an international conference. The model should be used for the automated diagnosis of epilepsy, Alzheimer's disease, ADHD, and Autism Spectrum Disorder (ASD) (Adeli, 2011).

2.4 Heart rate variable (HRV)

A well established method for the diagnosis of mental disorders is the measurement of heart rate variability. It can be measured by variation in the R-R interval of an electrocardiogram (ECG). For further reading we suggest reading (Berntson et al., 1997) and (Miyamoto et al., 1992). Billman has described a historical perspective of heart rate variability. Therefore, further reading of (Bilmann, 2011) is recommended. The H. Tsuji study showed that reduced heart rate variability has been reported to predict risk for subsequent mortality. (Tsuji et al., 1994).

Gary G. Berntson and John T. Cacioppo showed that stress has a clear impact on the autonomic nervous system. It can be seen through lowering the heart rate variable (HRV) (Berntson & Cacioppo, 2004). The Netherland study shows that depression is associated with significantly lowered heart rate and respiratory rate variability. However, this association appears to be mainly driven by the effect of antidepressants (Licht et al., 2008). The Andrew H. Kemp and associates study compares HRV in patients with a major depressive disorder and healthy control subjects and the HRV of patients with a major depressive disorder before and after treatment using meta-analysis. They concluded that depression without cardio vascular disease is associated with reduced HRV, which decreases with increasing depression severity and is most apparent with nonlinear measures of HRV. Critically, a variety of antidepressant treatments (other than Tricyclic antidepressants

[TCAs][2]) neither increases nor decreases HRV (Kemp et al., 2010). Another study concluded that adolescent female psychiatric patients with anxiety disorder and/ or major depression disorder had a reduced HRV compared with healthy controls. Medication with selective serotonin reuptake inhibitor (SSRI)[3] explained part of this difference (Blom et al., 2010). Chalmers and his team found that anxiety disorders are associated with reduced HRV; these findings are associated with a small-to-moderate effect size. These findings have important implications for future physical health and well-being of patients, highlighting a need for comprehensive cardiovascular risk reduction (Chalmers et al., 2014). Similar research findings are presented by (Miu et al., 2009) (Friedman & Thayer, 1998) (Thayer et al., 1996). One of the plausible preventions suggest that modest amounts of regular moderate-to-vigorous physical activity sufficient to slow the accelerating age-related decline in cardiorespiratory fitness during late middle-age has protective benefits against the onset of depression complaints in both men and women (Dishman et al., 2012). Gorman and Sloan published an observation of heart rate variability that depressed patients after myocardial infarction exhibit higher mortality rates compared with non-depressed patients. Men with "phobic anxiety" also have higher rates of sudden cardiac death and coronary artery disease than control populations. Although HRV is reduced, treatment with the selective serotonin reuptake inhibitor paroxetine normalises heart rate variability. Hence, there is potential for the treatment of psychiatric disorders to positively affect the development and course of cardiovascular disease (Gorman & Sloan, 2000).

2.5 Pulse – respiratory coupling

Some studies observe pulse in respiratory system separately. They analysed ECG and respiratory signal and search for possible connections with mental disorders (Licht et al., 2008). Widjaja and his college found that during stress, attention heart and respiratory rate is increased when compared to a resting condition (Widjaja et al., 2013).

A novel approach has been performed; instead of separately analysing cardio and respiratory data, the observation of the autonomic nervous system through cardiorespiratory coupling is used. Kralemann and his colleges show that the phase at which the cardiac beat is susceptible to respiratory drive and extract the respiratory-related component of heart rate variability (Kralemann et al., 2013). Möser and his team use a multidimensional approach to describe the autonomic

---

2   TCAs are chemical compounds used primarily as antidepressants.
3   SSRI is used as an antidepressant.

nervous system and conclude that we can observe a central events influence of this activity in coordination of physiological parameters only with peripheral autonomic activity measurement (Moser et al., 1995). There are still many ongoing research studies on this field and we can expect a breakthrough in this field of research.

3 Conclusion

Serious work has been done in the area of observation of the autonomic nervous system. Previous research describes many influencing factors that can interfere functions of the autonomic nervous system. Among them are mental disorders which have a serious proportion in human populations. In future studies, a systematical investigation of a lager group of subjects and patients must take place, so that we can have an automatic and non-subjective method to diagnose mental disorders. A huge gap is still needed to be overcome in the area of multiple mental disease diagnostics.

In the future, we expect that the diagnostics will rely more on the analysis of the data. In fact, modern biomedical experiments generate enormous amounts of data on various body parameters, among them cardio-respiratory data and HRV data. Using statistical analysis, data mining methods, and automated machine learning procedures, we could have faster and more accurate predictions of diagnosis and hopefully have some new progress in differential diagnosis. Those findings may find an important place in the medical practitioner toolbox. It is particularly important to initiate appropriate treatment when expert psychiatric assessments may not be available for many weeks.

4 Acknowledgments

We would like to thank Max Möser for consultation and the Faculty for Information Studies for their support. We also would like to thank the Ministry of Education, Science, and Sport and the Republic of Slovenia for their financial support.

5 References

Adeli, H., 2011. Automated EEG-based diagnosis of the neurological and psychiatric disorders. In *Systems, Signals and Image Processing (IWSSIP)*. Sarajevo, 2011. IEEE.

Albrecht, A.T. / Herrick, C., 2007. *100 questions and answers about bipolar (manic depressive) disorder*. Sudbury, Massachusetts: Jones and Bartlett Publishers.

Berntson, G.G./ et al., 1997. Psychophysiology. *Heart rate variability: origins, methods, and interpretive caveats*, November. pp. 623–648.

Berntson, G.G. / Cacioppo, J.T., 2004. *Dynamic electrocardiography*. 1st ed. New York: Futura publishing company Inc.

Bilmann, G.E., 2011. Frontiers in psysiology. *Heart rate variability - a historical perspective*, 29 November.

Blessingm, B. / Gibbins, I., 2008. *Scholarpedia*. [Online] Available at: http://www. scholarpedia.org/article/Autonomic_nervous_system [Accessed 9 October 2014].

Blom, H.E. / et al., 2010. Acta Paediatrica. *Heart rate variability (HRV) in adolescent females with anxiety disorders and major depressive disorder.*, 28 January. pp. 604–611.

Brading, A., 1999. *Autonomic Nervous System and Its Effectors*. Hoboken, Ney Jersy: Wiley-Blackwell.

Bremner, J.D. / et.al., 2000. *The American Journal of Psychiatry*. [Online] Available at: http://journals.psychiatryonline.org/article.aspx?articleid=173909 [Accessed 15 October 2014].

Chalmers, J.A./ Quintana, D.S./ Abbott, M.J.-A. / Kemp, A.H., 2014. Frontiers in Psychiatry. *Anxiety Disorders are Associated with Reduced Heart Rate Variability: A Meta-Analysis.*, 11 June. pp. 1–11.

Dishman, R.K./ et al., 2012. American journal of preventive medicine. *Decline in Cardiorespiratory Fitness and Odds of Incident Depression*, 1 October. pp. 361–368.

Friedman, B.H. / Thayer, J.F., 1998. Journal of Psychosomatic Research. *Autonomic balance revisited: Panic anxiety and heart rate variability*, 1 January. pp. 133–151.

Gilbertson, M., 2002. *US National Library of Medicine National Institutes of Health*. [Online] Available at: http://www.ncbi.nlm.nih.gov/pubmed/12379862 [Accessed 15 October 2014].

Goldberg, J., 2014. *WebMD Medical Reference*. [Online] Available at: http://www. webmd.com/anxiety-panic/guide/mental-health-making-diagnosis [Accessed 12 October 2014].

Goldberg, D.P. / et.al., 1997. Psychological Medicine. *The validity of two versions of the GHQ in the WHO study of mental illness in general health care*.

Gorman, J.M. / Sloan, R.P., 2000. American Heart Journal. *Heart rate variability in depressive and anxiety disorders*, 1 October. pp. 77–83.

Kemp, A.H. / et al., 2010. Biological Psychiatry. *Impact of Depression and Anti-depressant Treatment on Heart Rate Variability: A Review and Meta-Analysis*, 1 June. pp. 1067–1074.

Khodayari-Rostamabad, A. / et al., 2010. Diagnosis of psychiatric disorders using EEG data and employing a statistical decision model. In *Piscataway, NJ:IEEE Service Center*. Buenos Aires, 2010. EMBS Annual International Conference of the IEEE.

Kralemann, B./ et al., 2013. Nature communications. *In vivo cardiac phase response curve elucidates human respiratory heart rate variability*, 2 September. pp. 1–9.

Licht, C.M.M. /et al., 2008. Archives of General Psychiatry. *Association between major depressive disorder and heart rate variability in the Netherlands Study of Depression and Anxiety (NESDA)*, 1 December. pp. 1358–1367.

Mai, J.K. / Paxinos, G., 2012. *The Human Nervous System*. 3rd ed. London: Academic Press.

Makin, S., 2013. Scientific Americat Ô. *Can Brain Scans Diagnose Mental Illness?*.

Marlow, K., 2012. *The Superhuman Mind*. [Online] Available at: http://www.psychologytoday.com/blog/the-superhuman-mind/201212/diagnostic-breakthrough-mental-illness [Accessed 13 October 2014].

Miu, A.C./ Heilman, R.M. / Miclea, M., 2009. Autonomic Neuroscience: Basic and Clinical. *Reduced heart rate variability and vagal tone in anxiety: trait versus state, and the effects of autogenic training.*, 28 June. pp. 99–103.

Miyamoto, M./ Ichimaru, Y. / Katayama, S., 1992. Nihon rinsho. *Heart rate variability*, April. pp. 717–722.

Moser, M./ et al., 1995. Biological Rhythm Research. *Phase- and Frequency Coordination of Cardiac and Respiratory Function*, pp. 100–111.

Plesman, J., 2011. *Hypoglycemic Health Association of Australia*. [Online] Available at: http://www.hypoglycemia.asn.au/2011/anxiety-and-the-autonomic-nervous-system/ [Accessed 12 October 2014].

Thayer, J.F./ Friedman, B.H. / Borkovec, T.D., 1996. Biological Psychiatry. *Autonomic characteristics of generalized anxiety disorder and worry*, 15 February. pp. 255–266.

Tsuji, H./ et al., 1994. Circulation. *Reduced heart rate variability and mortality risk in an elderly cohort. The Framingham Heart Study.*, 1 Augustus. pp. 878–883.

Tsuji, H./ et al., 1994. Circulation. *Impact of reduced heart variability on risk of cardiac events The Framingham Heart Study.*

Widjaja, D./ Orini, M./ Vlemincx, E. / Van Huffel, S., 2013. Computational and Mathematical Methods in Medicine. *Cardiorespiratory Dynamic Response to Mental Stress: A Multivariate Time-Frequency Analysis*, 18 October. pp. 1–12.

Dragana Miljkovic[1], Kristina Gruden[2], Nada Lavrač[1,3,4]
[1]Jožef Stefan Institute, Jamova 39, Ljubljana, Slovenia
[2]National Institute of Biology, Ljubljana, Slovenia
[3]Jožef Stefan International Postgraduate School, Jamova 39, Ljubljana, Slovenia
[4]University of Nova Gorica, Vipavska 13, Nova Gorica, Slovenia
{dragana.miljkovic, nada.lavrac}@ijs.si
kristina.gruden@nib.si

Improving biological models with experts knowledge and literature

Abstract: A conventional way of manual construction of a dynamic model can be accompanied by additional steps which can speed up and enhance model construction. In our work, we have developed a new methodology for constructing biological models based on domain knowledge and literature. We have applied this methodology on the construction of the defence response model in plants. Using solely manual knowledge, engineering was not feasible due to the complexity of a plant defence system. For this reason, the laborious work of manual model creation was complemented by additional automatised and semi-automatised steps. These steps address the model structure, which was enhanced by means of natural language processing techniques, as well as model dynamics, where the parameter optimisation was guided by constraints defined by the domain experts.

Keywords: systems biology, plant defence, triplet extraction, natural language processing, evolutionary algorithms

1 Introduction

In biological sciences, the growth of experimental data is not uniform for different types of biological mechanisms. Therefore, some biological mechanisms still have few datasets available. The paper describes a novel methodology for the construction of biological models by eliciting the relevant knowledge from literature and domain experts. The methodology has been applied to build the model of defence response in plants against virus attacks. The developed plant defence model consists of three sub-models: salicylic acid (SA), jasmonic acid (JA), and ethylene (ET) sub-model, since SA, JA and ET are the most important components of the defence mechanism in plants. The methodology addresses two aspects of biological model construction: the model structure and the model dynamics.

To construct the model structure, different information sources can be used. Given that most of the human biological knowledge is stored in the silos of biological literature, retrieving information from the literature is useful when building the biological models. State-of-the-art technologies enable information extraction from scientific texts in an automated way by means of text processing techniques, based on the advances in the area of natural language processing (NLP) of biology texts. NLP is used in systems biology to generate the model structures or to enhance existing ones. The most common NLP approaches can be classified into three categories (Cohen, Hunter 2008): rule-based approaches, machine-learning approaches, and co-occurrence-based approaches. Examples of rule-based systems include GeneWays (Rzhetsky et al. 2004), Chilibot (Chen, Sharp 2004), etc. There are also combined methods, including co-occurrence-based approaches, such as Suiseki developed by Blaschke and Valencia (2002). A wide range of machine learning techniques is used for relations extraction in systems biology, like the Naive Bayes classifier (Craven, Kumlien 1999), Support Vector Machines (Donaldson et al. 2003), etc. Due to the small amount of existing quantitative data, mathematical optimisation methods for parameter optimisation were recently employed in systems biology. Various local deterministic optimisation techniques, like the Levenberg-Marquardt algorithm (Levenberg 1944), and stochastic approaches, like Genetic Algorithms (Mitchell 1966) and Evolutionary Algorithms (Eiben, Smith 2003), are applied in systems biology.

The goal of this paper is to present the developed methodology which enables building complex biological models without or with scarce experimental data. The methodology consists of several steps, where the standard approach to the construction of dynamic models is enhanced with the following methods: a method for model structure revision by means of natural language processing techniques, a method for incremental model structure revision, and a method for automatic optimisation of model parameters guided by the expert knowledge in the form of constraints.

2 Materials and methods

An overview diagram of the iterative model construction process is shown in Figure 1. The most relevant details of every step in this construction process are explained in the following subsections.

Figure 1: *A schema of the developed methodology for the plant defence model construction*

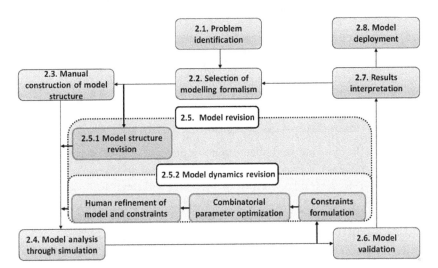

2.1 Problem identification

In this step, we have defined the requirements for the plant defence model in collaboration with the experts from the National Institute of Biology in Ljubljana, Slovenia. The goals of model development are the following: a) better understanding of the biological mechanism on the system level and b) prediction of experimental results with the aim to detect crucial reactions in the plant defence process, predict the final defence response when some genes are silenced, and discover new connections/interactions.

2.2 Selection of modelling formalism

The decision of which modelling formalism to select depends on the requirements defined by the domain experts, currently available knowledge, and open issues related to the plant defence response. In the plant defence research field, there is few experimental data and on the other hand, there is a lot of domain knowledge related to the modelling of the plant defence mechanism that has not yet been formalised and systematised. For these reasons, we first present the model structure in the form of a directed edge-labelled graph[1], which is a common way

---

1   Directed edge-labelled graph of the plant defence model structure represents biological components as nodes and biological reactions as vertices between them.

to present biological networks in systems biology and which is, at the same time, intuitive for the domain experts.

2.3 Manual construction of model structure

The initial plant defence model structure was constructed manually by considering knowledge from the literature, different biological databases such as TAIR (Swarbreck et al. 2008) and KEGG (Kanehisa, Goto 2000), and domain experts.

2.4 Model analysis through simulation

Analysis of the dynamic model behaviour was performed through iterative simulations of the manually constructed plant defence model. We selected Hybrid Functional Petri net (HFPN) formalism to simplify and simulate the plant defence model, whose structure was developed initially in the graph form. The simulation was initially performed by Cell Illustrator (CI) (Nagasaki et al. 2004), which implements the HFPN formalism. The simulator outputs time series curves of the dynamic behaviour of components of interest.

2.5 Model revision

The model revision process includes the revision of model structure and its dynamics.

2.5.1 Model structure revision

The process of fusing expert knowledge and manually obtained information from the literature to build the plant defence model structure turns out to be time-consuming, non-systematic, and error-prone. Therefore, we have introduced one additional step that enhances the manually built model structure: extraction of relations between biological components from the literature, using the natural language processing approach. We have developed the Bio3graph tool (Miljkovic et al. 2012) that searches literature for the relations between the biological components and outputs a graph of triplets in the form (component1, relation, component2). Moreover, we have also developed an incremental version of the Bio3graph tool, which updates the network structure with new sets of triplets having an initial model as a baseline.

2.5.2 Model dynamics revision

This process consists of three steps:

a. Constraints formulation. The constraints are mathematical expressions defined by the domain experts. They represent the rules of how the simulation output curves of certain biological components should look like. The purpose of constraints is to guide and speed up the parameter optimisation search by limiting the parameter search space.
b. Combinatorial parameter optimisation. We used the differential evolution algorithm for parameter optimisation developed by Filipic and Depolli (2009). This algorithm performs a population-based search that optimises the problem by iteratively trying to improve a candidate solution with regard to a measure of quality.
c. Human refinement of model and constraints. If dynamic behaviour of the curves does not fit the experts' expectations, the model structure and the constraint setting are modified.

2.6 Model validation

The validation of the simulation results is based mainly on the judgement of the domain experts to ensure that the model is close to the real-life system. However, the simulation outputs can also be validated with the validation method for non-observable systems, such as parameter sensitivity analysis.

2.7 Results interpretation

The simulation results are interpreted by the biology experts. The output curves of the simulation allow for qualitative conclusions regarding the dynamic behaviour of the model. For example, by comparing the growth of certain curves, we can determine the components that contribute the most to the plant defence response.

2.8 Model deployment

The finalised version of the model will be used at National Institute of Biology in Ljubljana (and hopefully in broader scientific community) to assist the experimental design by generating hypotheses about how the plant defence response will react when particular genes are silenced.

3 Results and discussion

To successfully build the plant defence model, we have developed three semi-automated methods to enhance manual model construction: a method for model structure revision by means of natural language processing techniques, a method for incremental model structure revision, and a method for automatic optimisation of model parameters guided by the expert knowledge in the form of constraints. Using these implemented methods, the enhanced model of the defence response in plants was constructed.

3.1 Initial plant defence model

As a first step, the plant defence model structure is compiled manually in the form of a directed edge-labelled graph and visualised with the Biomine visualisation engine (Eronen, Toivonnen 2012). Reactions are presented as graph arcs labelled with one of three reaction types: activation (A), inhibition (I), and binding, and components as graph nodes connected with arcs to each other. We present the model structure in the form of a directed edge-labelled graph in Figure 2A, consisting of 175 nodes and 387 reactions.

The model structure in Figure 2A is not simple for an expert to manually determine all the model parameters and to grasp and synchronise the dynamic behaviour of all feedback loops which are modelled. Since this model structure was too complex to be used directly for simulation purposes, we have simplified it and converted for simulation with the CI software (Nagasaki et al. 2004). The software facilitates the easy building of the network structure, has a graphical editor that has drawing capabilities, and allows biologists to model different biological networks and simulate the dynamic interactions between the biological components. The final result of a model prepared for the simulation is shown in Figure 2B.

The developed model structure, shown in the form of graph in Figure 2A, contains more detailed information compared to the structural model of the subsets of plant defence (Staswick, Tiryaki 2004), having 175 components and 387 reactions in total.

The model structure transformed to the HFPN presentation is prepared for the analysis of model dynamics. A slightly reduced structure (Figure 2B), compared to the directed edge-labelled graph of Figure 2A, contains 99 components and 68 reactions in total. The structure of one of the first simulation models of the plant defence response (Genoud et al. 2001), containing 18 biological entities and 12 Boolean operators, is less complex than the model structure shown in Figure 2B.

Figure 2: Plant defence model. A) Model structure in the form of a directed edge-labelled graph. B) Simplified model, developed according to the HFPN formalism for simulation purposes

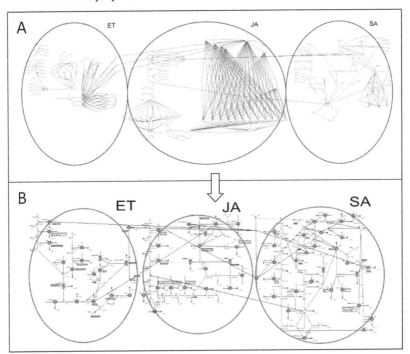

3.2 Method for constraint-guided automatic optimisation of model parameters

As parameter hand tuning of the merged plant defence model (Figure 2B) was not feasible due to the model's complexity, it was necessary to support this process by means of parameter optimisation. We selected the most studied SA sub-model, presented in the form of HFPN formalism, and refined it through several iterative optimisations.

A simple biological reaction represented according to the HFPN formalism is presented in Figure 3. Behind the graphical presentation (Figure 3A) is the set of differential equations (Figure 3B). The differential evolutionary algorithm (Filipič, Depolli 2009) searches for optimal values of reaction rates (k) and inhibition thresholds (t_{thr}) (Figure 3B) with the criteria that constraints have to be minimally violated. The constraints are collected from domain experts in the form of relationships between biological components and are formalised to narrow the parameter search space.

The following five types of constraints are defined: (1) inequality relationship between quantities of molecules, (2) growth rate of the molecules' quantities (i. e. the quantity of molecule one grows faster than that of molecule two), (3) curve shape (i. e. it starts at some initial level, reaches a maximum, and then drops back to an approximately similar level), (4) minimal amplitude and minimal growth of a curve, and (5) temporal sequence in curve maxima.

We refined the SA sub-model through three iterative steps until the domain experts were satisfied with the simulation outputs and the criteria function of the optimisation algorithm was minimal (Miljkovic et al. 2014). The analogous process can be applied to the JA and ET sub-models.

Figure 3: A simple biological reaction presented with HFPN formalism. A) Graphical presentation of reaction within HFPN formalism. B) Differential equations of reaction which are related to the graphical presentation in A).

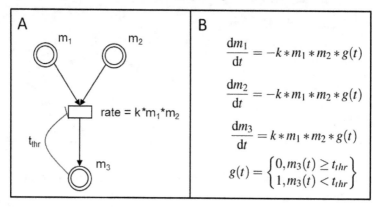

3.3 Method for model structure revision

The process of fusing expert knowledge and manually acquired information can be enhanced by automated methods of relation extraction from literature, which are recently popular in systems biology. The most common relation extraction methods from texts are based on natural language processing techniques. Ideally, the output of such relation extraction methods is a graph of biological components and relations between them.

The proposed Bio3graph methodology was developed with the purpose of automated information extraction from biological literature, aimed at complementing the manually developed plant defence model structure. An integral part of this methodology is a domain specific vocabulary that is composed of two parts: a list of components and a list of reactions together with their synonyms. The basis for the vocabulary was the list of 175 components and three reaction types defined when building the manual plant defence model structure. The components vocabulary consists of their short names, gene identifiers and synonyms, as annotated in TAIR (Swarbreck et al. 2008). As several components included in the manual PDS model structure are still not fully identified, they were labelled as X in the PDS model structure and were not included in the vocabulary. In addition, most of the complexes were not included in the vocabulary except for the SCF complex. Consequently, the list of 153 biological components in the vocabulary contains fewer components than the vocabulary of 175 components used for the manual PDS model structure construction. Furthermore, the vocabulary for the reaction types was developed, containing synonyms for the three reaction types: activation, inhibition, and binding. The Bio3graph methodology consists of a series of text mining, information extraction, graph construction, and graph visualisation steps, offering reusability, repeatability, and extension with additional components (see Bio3graph box in Figure 4).

The methodology is implemented as a workflow in the Orange4WS (Podpečan et al. 2012) workflow construction and execution environment. The input to the Bio3graph workflow is the collection of biological full text articles, obtained through a user-defined keyword-based search of the PMC database, accessible at: www.ncbi.nlm.nih.gov/pmc/. The output of the workflow is a graph of triplets, automatically extracted from the articles.

We extracted 137 new triplets from 9,586 scientific articles and revised the initial manually constructed model structure into a single graph consisting of 175 components and 524 relations (Miljkovic et al. 2012).

3.4 Method for incremental model structure revision

After biological models are published, they do not tend to be quickly re-examined. However, there is a constant flow of new knowledge related to the biological mechanism. The fast updates of the biological model structures are possible with the incremental approach based on the relation extraction from literature.

The work on incremental revision of biological models is based on the Bio3graph approach (described in previous subsection), which allows for auto-mated extraction of biological relations in the form of triplets from the literature. Apart from the steps of Bio3graph tool, its incremental extension implements additional steps: literature retrieval, graph merging, redundant relation removal, and colour reset. We define the inputs to the incremental extension as follows

(Figure 4). The existing model structure, which is the subject of incremental en-hancement, is called the "Initial graph" and the result of Bio3graph is called the "Triplet graph". The incremental extension of Bio3graph produces two outputs: "Incremented graph", a result of merging the Initial and the Triplet graph, and "Filtered incremented graph", a result of redundant transitive relation removal from the "Incremented graph".

Figure 4: Scheme of the methodology for incremental construction of biological networks using information extraction from literature.

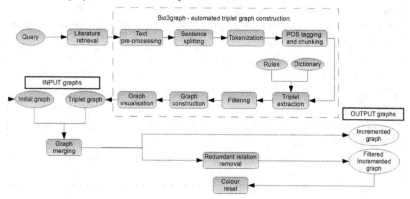

4 Acknowledgements

This work was financed AD Futura Agency, Slovenian Research Agency grants P4-0165, J4-2228, J4-4165, J2-5478 and P2-0103.

5 References

Blaschke Christian / Valencia Alfonso: The frame-based module of the Suiseki information extraction system. IEEE Intelligent Systems 17, 2002, pp. 14–20.

Chen Hao / Sharp Burt M: Content-rich biological network constructed by min-ing pubmed abstracts. Bmc Bioinformatics 5, 2004.

Cohen Kevin B / Hunter Lawrence: Getting started in text mining. Plos Compu-tational Biology 4(1), 2008.

Craven Mark / Kumlien Johan: Constructing biological knowledge bases by ex-tracting information from text sources. Proceedings of International Confer-ence on Intelligent Systems for Molecular Biology, 1999, pp. 77–86.

Donaldson Ian / Martin Joel / de Bruijn Berry / Wolting Cheryl / Lay Vicki / Tuekam Brigitte / Zhang Shudong / Baskin Berivan / Bader Garry D / Michal-ickova Katerina / Pawson Tony / Hogue Christopher WV: PreBIND and

Textomy – mining the biomedical literature for protein-protein interactions using a support vector machine. BMC Bioinformatics 4, 2003.

Eiben Agoston E / Smith James E: Introduction to Evolutionary Computing. Springer-Verlag, Berlin, 2003.

Eronen Lauri / Toivonnen Hannu: Biomine: predicting links between biological entities using network models of heterogeneous databases. BMC Bioinformatics 13, 2012.

Filipič Bogdan / Depolli Matjaž: Parallel evolutionary computation framework for single- and multiobjective optimization. Parallel Computing 217–240, 2009.

Genoud Thierry / Santa Cruz Marcela B. Trevino / Metraux Jean-Pierre: Numeric simulation of plant signaling networks. Plant Physiology 126, 2001, pp. 1430–1437.

Kanehisa Minoru / Goto Susumu: KEGG: Kyoto Encyclopedia of Genes and Genomes. Nucleic Acids Research 28, 2000, pp. 27–30.

Levenberg Kenneth: A method for the solution of certain non-linear problems in least squares. Quarterly of Applied Mathematics 2, 1944, pp. 164–168.

Miljkovic Dragana / Depolli Matjaž / Stare Tjaša / Mozetič Igor / Petek Marko / Gruden Kristina / Lavrač Nada: Plant defence model revisions through iterative minimisation of constraint violations. International Journal of Computational Biology and Drug Design 9(7), 2014, pp. 61–79.

Miljkovic Dragana / Stare Tjaša / Mozetič Igor / Podpečan Vid / Petek Marko / Witek Kamil / Dermastia Marina / Lavrač Nada / Gruden Kristina: Signalling network construction for modelling plant defence response. PLOS ONE 7, 2012.

Mitchell Melanie: An Introduction to Genetic Algorithms. MIT Press, Cambridge, MA, 1996.

Nagasaki Masao / Doi Atsushi / Matsuno Hiroshi / Satoru Miyano: Genomic Object Net: I. A platform for modeling and simulating biopathways. Applied bioinformatics 2, 2004, pp. 181–184.

Podpečan Vid / Zemenova Monika / Lavrač Nada: Orange4WS Environment for Service-Oriented Data Mining. The Computer Journal 55, 2012, pp. 82–98.

Rzhetsky Andrey / Iossifov Ivan / Koike Tomohiro / Krauthammer Michael / Kra Pauline / Morris Mitzi / Yu Hong / Duboue Pablo Ariel / Weng Wubin / Wilbur W. John / Hatzivassiloglou Vasileios / Friedman Carol: Geneways: a system for extracting, analyzing, visualizing, and integrating molecular pathway data. Journal of Biomedical Informatics 37, 2004, pp. 43–53.

Staswick Paul E / Tiryaki Iskender: The oxylipin signal jasmonic acid is activated by an enzyme that conjugates it to isoleucine in arabidopsis. Plant Cell 16, 2004, pp. 2117–2127.

Swarbreck David / Wilks Christopher / Lamesch Philippe / Berardini Tanya Z / Garcia-Hernandez Margarita / Foerster Hartmut / Li Donghui / Meyer Tom / Muller Robert / Ploetz Larry / Radenbaugh Amie / Singh Shanker / Swing Vanessa / Tissier Christophe / Zhang Peifen / Huala Eva: The Arabidopsis Information Resource (TAIR): Gene structure and function annotation. Nucleic Acids Research 36, 2008, D1009-D1014.